no more myths

no more more myths

Real Facts to Answer Common Misbeliefs About Pet Problems

Written and Illustrated by

Stefanie Schwartz, BSc, DVM, Msc
Veterinarian, Pet Behavior Consultant

 HOWELL BOOK HOUSE

Howell Book House
MACMILLAN
A Simon & Schuster Macmillan Company
1633 Broadway
New York, NY 10019

MACMILLAN is a registered trademark of Macmillan, Inc.
Library of Congress Cataloging-in-Publication Data

Schwartz, Stefanie.
    No more myths : real facts to answer common misbeliefs about pet
problems / written and illustrated by Stefanie Schwartz.
        p.   cm.
    ISBN 0-87605-692-3
    1. Dogs—Behavior.   2. Cats—Behavior.   3. Pets—Behavior.
4. Dogs—Psychology.   5. Cats—Psychology.   6. Pets—Psychology.
I. Title
SF433.S386   1996
636.7' 0887—dc20   96–11243
                                                            CIP

Manufactured in the United States of America
10 9 8 7 6 5 4 3 2

Book Design by MIGUEL SANCHEZ

This book is dedicated to my parents

Contents

Author's Preface   XV

1. General Misbeliefs About Cats and Dogs   1

Dogs bond more strongly to their owners, and cats bond more strongly to
their home territory • Animals don't dream • Dogs or cats with blue eyes
are always deaf • Cats and dogs see in black and white only • To keep your
pet's nose black (not pink), feed it from a ceramic bowl • A dog or cat's
name can never be changed • You can predict how big a puppy or kitten
will grow by paw size • Cats and dogs have short attention spans

2. General Misbeliefs About Cats   9

Cats have nine lives • Cats are not demonstrative pets • Cats purr only when
they are happy • Cats wag their tails only when they are angry • Cats need
catnip as a part of a healthy diet • All cats love catnip • Cats are always good
hunters • Cats must roam outdoors to live happy lives (It is cruel to
confine a cat indoors) • An outdoor cat can never adapt to confinement as
a house cat • Cats always land on their feet • Cats see better at night than
during the day • Trimming a cat's whiskers will cause blindness • Abundant hair in the ears indicates the cat is male • Stripes on the tail indicate a
cat is male • Striped markings indicate a cat is descended from tigers • Black
cats are aggressive and unlucky

3.   General Misbeliefs About Dogs   21

Dogs are direct descendants of wolves • Dogs do not need to be walked on a leash if they have access to a fenced yard • A dog can be left alone in a fenced yard • It is okay to leave a dog tethered in your yard • "Unseen fences" (that deliver painful shocks or high-frequency sound) are foolproof • Purebred dogs have better temperaments than mixed breeds or "mutts" • Purebred dogs are better pets than mixed breeds • During car rides, dogs should be allowed to have their heads out the window for fresh air and to enjoy the drive • It is safe for a dog to ride in the back of a pick-up truck • AKC registration papers certify a dog's quality • Dogs that run away from home can find their own way back • Dogs run away from home because they are ungrateful, angry or dissatisfied with their food • You should punish your dog for running away • Dogs have no preference for using a right or left paw • On a hot day, your dog is better off in the car than at home • Dogs are meant to live outside in cold weather • A dog's instinct is to please all humans

4.   Pet Selection   33

Selecting a pet is an emotional and spontaneous decision • The most important consideration when selecting a pet is how it looks • Your finances are irrelevant to acquiring a pet • No special skills or experience are required to raise a dog or cat successfully • The "pick of the litter" is the biggest one • The "runt" will be sickly and die • Hunting dogs salivate more than other dogs, and this is part of having a "soft mouth" • Pet behavior problems are usually due to loneliness, so adding a second pet is the solution • When adding another pet, a male and female always get along • If aggression occurs between a newly introduced dog and the resident dog, they will never get along well • If aggression occurs between a newly introduced cat and the resident cat, they will never get along well • The best way to introduce two pets is to confine them together in a room and check on them later when things are quiet • If you do not own a house with a fenced yard, you should not have a dog • Large breed dogs should not live in apartments • Smaller dogs are less aggressive and make better pets for families with young children • Cats are women's pets • Cats make good pets for busy people because they require no attention and little care • Siamese are more vocal and intelligent than other cats • Intelligence in dogs varies according to breed • Border Collies are the smartest breed • Afghan Hounds are the least intelligent breed • Chocolate Labrador Retrievers are easier to train than Labs of other colors • Sheepdogs relate to their owners as they would to their flock • Some dog and cat breeds are "hypoallergenic" • If you are allergic, your only option is to place your pet in another home • The age at which a pet is acquired is not important • Pet store puppies are no different from dogs bred by breeders

5. Pets and Children 57

The best way to teach the facts of life is to let your children observe the birth of puppies or kittens • Children will learn to be responsible by caring for a pet • Children are naturally kind to other animals • A good dog or cat has endless patience with children • Children should grow up with pets from a young age • A woman who is pregnant or considering having a child must avoid all contact with cats • It is safe for children to pet a dog who is walking with the owners • If your pet has never been aggressive toward children, it will not endanger your infant in any way • If kids are fighting, a good dog will jump on them to break it up • If a dog seems harmless to an infant, there should be no problems between them as the child grows • Male dogs are more likely than females to become aggressive toward a new baby

6. Obedience 67

Obedience training should begin when your puppy is at least six months old or after neutering • Physical force is the best way to gain a pet's respect • A food reward must always be given during training • Punishment must be painful or cause fear to be effective • Punishment or discipline is more effective than reward in modifying undesirable behavior • Striking your pet with a rolled up newspaper is effective discipline • Obedience training is unpleasant for dogs • Small dogs do not require Obedience training • A good watchdog or hunting dog must be kept outside at all times • Deaf dogs cannot be Obedience trained • Cats cannot be trained to do tricks or respond to commands • The five basic Obedience commands are "Sit down," "Lay down," "No," "Come here" and "Get down"

7. Aggression 79

Do not disturb a sleeping dog • Dogs will never bite the hand that feeds them • Female dogs make better watchdogs • Play-biting is harmless and pups soon outgrow it • A dog should be encouraged to play rough and wild games in order to behave more calmly later • Dogs are always aggressive toward postal carriers • Mounting behavior indicates a dog is sexually aroused • Mounting behavior is not seen in neutered dogs • Dominance aggression is always expressed by mounting • Mounting behavior in female dogs indicates sexual confusion • Mounting behavior in dogs begins at puberty (about six months of age) • Staring prevents an aggressively aroused dog from harming you • A dog with a wagging tail is friendly and will not bite • Dogs always growl before they bite • All American Pit Bull Terriers are extremely aggressive • All Golden Retrievers are gentle and trustworthy • A pet that becomes aggressive or excessively fearful at the veterinary clinic must have been abused there (or at another clinic) in the past • A pet that is aggressive toward men has been beaten by one • A dog that becomes aggressive if disturbed during feeding was probably starved as a puppy • Possessive aggression (guarding) occurs only with food and indicates the dog's dominance • A pet that is aggressive toward the owner is ungrateful • A dog that becomes aggressive toward the owner's mate is exhibiting jealousy • A pet that suddenly becomes aggressive is having a seizure • A pet that suddenly becomes aggressive probably has rabies • Calico cats are more aggressive than cats of other coat colors • A cat will become aggressive or more likely to bite if declawed • Petting or brushing a pet against the direction of coat growth will cause aggression • Dog fights and cat fights are to the death • All dogs hate cats • All cats hate dogs and will scratch out their eyes

8. Destructiveness   103

Dogs that dig are looking for buried bones • Digging occurs only outdoors • A pet that is destructive when left alone is acting out of anger or spite • Destructive behavior is always due to separation anxiety • Destructiveness is the only sign of separation anxiety • Cats use a scratch post to sharpen their claws • Cats scratch where they shouldn't because of owner-directed anger • Cats instinctively use scratch posts and do not have to be taught to use them • Scratch posts must be treated with catnip to be effective • It is always cruel to declaw an indoor cat • Trimming a cat's nails will increase scratching at undesirable locations • Cats that suck or chew on cloth surfaces were weaned too early

9. Elimination Problems   115

Paper training is the best method to house train a puppy • The best way to punish house soiling is to stick your pet's nose in its own waste • You can tell there has been an "accident" because your dog "looks guilty" • When urinating, female dogs always squat and males always stand and lift a leg •

Male dogs need to urinate more frequently than females • A dog that leaks urine in greeting has defective sphincters • Unlike cats, dogs do not mark their territory by voiding indoors • Dogs eat stool because they have a nutritional deficit or abnormal temperament • Cats do not eat their own feces because they are smarter than dogs • Coprophagia is corrected by treating stools with a chemical or distasteful substance or by changing the dog's diet • Starving your dog will teach bowel control • "Scooting" is usually a sign of internal parasites • A pet that voids indoors when you are home is acting out of spite • A pet that voids indoors when left alone is acting out of spite • Restrict or remove water to stop inappropriate urination • Urine marking is prevented by neutering a cat before six months of age • A litter box needs changing only once or twice each month • A cat that does not cover litter box waste is abnormal • Only male cats suffer from Feline Urological Syndrome (*FUS, cystitis, urinary crystals*)

10. Sex-Related Problems   131

Females should not be neutered before the first heat • A female dog or cat should have at least one "heat" or litter to be a better pet • Male dogs or cats should not be neutered because sexual fulfillment is necessary for them to be happy • All bitches or queens instinctively know how to deliver and care for their young • Injury inflicted on offspring by their dam is always accidental • Male cats or dogs recognize their own offspring and would never harm them • Males without prior experience always know how to mate • Neutering will make a male or female dog less protective of you and your home • Neutering a male dog or cat is the same as a vasectomy performed on a man • A "spayed" female retains her ovaries, but the uterus is removed • "False" pregnancy in a female is normal and indicates a desire to give birth • Female cats will mate with any male cat, and female dogs will mate with any male dog • Dogs that have mated must be pulled apart or they will remain stuck together permanently • It takes six months after neutering for a male's behavior to change • Both males and females come into "heat" around six months of age • Female dogs menstruate just like women • The best way to tell if a bitch is in estrus is to watch for vaginal bleeding • Female cats come into estrus only once or twice a year • Neutering will make a pet fat and lazy • Neutering will make a pet mean • Neutering will stop unwanted aggression in males and females • Neutering a male cat will create a predisposition to urinary blockages • Neutering will prevent urine spraying in male cats • Mounting behavior indicates a dog is sexually aroused • Male dogs and cats do not have nipples • There are no sexually transmitted diseases in pets • Males must be neutered early or they will learn to masturbate

11. Miscellaneous Behavior Problems   149

Dogs howl because they are unhappy • Dogs bark because they have something to say • Pets must be sedated for travel • Crate training is the best
way to raise a dog • A dog that rolls over in greeting has been abused
• Hyperactive pups require sedatives or psychoactive drugs • Pets eventually outgrow a fear of thunder or other loud noises • A pet that is fearful of
men has been beaten by one • Pets that investigate the "private parts"
of other dogs and people are abnormal • Jumping on people is a dog's way
of showing affection • Cat toys are only for cats, and dog toys are only for
dogs • Dogs and cats will outgrow car sickness • Pets should never be
boarded because they will stop eating and might die • Pets should never
be kenneled longer than one or two weeks because they will think they have
been abandoned • Cats do better during separations because they do not
love their owners • All boarding kennels are alike

12. Health   163

Pets do not suffer from the same diseases as people • Medication prescribed
for people may be used for pets with the same symptoms • Animals heal
themselves by licking • A warm, dry nose means your pet is ill • A cold and
wet nose means your pet is healthy • Vaccination is important only if your
pet goes outside or has direct contact with other animals • Vaccinations guarantee your pet will never get sick • The distemper vaccine prevents dogs
and cats from biting • Intestinal parasites affect only young pets • Intestinal parasites always cause diarrhea and vomiting • Parasitic worms can
always be identified by looking at them • Heartworm is detected by analyzing stool samples • Cats do not get heartworm • The hair in a dog's ears
should always be plucked • Certain dog breeds require cropped ears or
docked tails for health reasons • Dogs and cats eat lawn grass or other plants
because they need to vomit • Sneezing is a sign of allergy • Colds are passed
between people and pets • Brown discharge from a pet's eyes is a sign of
infection • Mange in dogs is always contagious • Ringworm in people is always caused by contact with cats • Cats cause AIDS (Acquired Immune
Deficiency Syndrome) • Teeth and gum problems do not affect a pet's general health • Bad breath is a sign of dental problems • Veterinarians are not
"real doctors" • Veterinary medicine is not as advanced as human medicine

13. Nutrition   181

Pets instinctively eat what they need to balance their diet • Cats and dogs
only eat what they need, so food can be available constantly • Adult cats

and dogs should only eat one daily meal • More female cats become overweight compared to male cats • The eating habits of pet owners do not affect the eating habits of their pets • Cats should be allowed continual access to food • Cats will not eat anything outdoors if they are well fed at home • Male cats should not eat dry cat food • Pets know when a food is not safe to eat • Strictly vegetarian diets are healthy for dogs and cats • All pet foods are alike—the best brand is the cheapest • Pet food labels reliably determine your pet's daily food intake • Pet foods labeled as "natural" or "health food" are better quality • Dry food prevents dental problems • Dogs must chew on bones as part of their diet • Chewing bones is good for your pet • Excessive water intake should not be permitted • Cats must drink milk because they do not drink water • Cats require milk in their diet • Drinking milk causes worms • A pet's diet must be varied to prevent boredom • Table scraps are important for a balanced diet • Adding brewer's yeast or garlic to your pet's diet will prevent fleas • Raw eggs ensure a healthy and shiny coat • Raw fish is healthy for cats • An exclusive diet of cooked or raw meat (or liver) is good for pets • A home-cooked meal is better than commercially prepared diets • A teaspoon of cooking oil ensures a shiny coat • Cooking oil will relieve constipation • Shedding hair can be prevented by medication or nutritional supplements • Shedding hair can be prevented by special sprays or frequent bathing • Your pet cannot develop an intolerance or allergy to food it has been eating for a long time • Dogs eat stool because they have a nutritional deficit or abnormal temperament • Pets eat lawn grass and other plants because they have a nutritional deficit • Salivating (drooling) means your pet is hungry • Neutering your pet will make it fat and lazy

14. Grooming   203

Healthy pets have no body odor • Dogs and cats, especially white ones, must be bathed at least weekly • Trimming the hair away from a dog's eyes will cause blindness • Cats do not need to be brushed • Only long-haired cats suffer from hairballs • A cat's tongue is like sandpaper • Short-haired cats and short-coated dogs do not need to be brushed or combed • Fleas always make a dog or cat scratch • Flea collars provide complete protection • New "flea pills" completely replace traditional treatments • All ticks carry Lyme disease • Dogs that have been sprayed by a skunk cannot be deodorized • Petting or brushing against the direction of coat growth will cause aggression • Pets have no dental problems • You can't brush your pet's teeth • Pets do not get dental cavities (*caries*) • Eating dry food takes care of a pet's basic dental care • Nothing can be done to treat a pet's tooth or gum problem • Chewing bones on a regular basis prevents/treats dental tartar • Nail-biting in dogs means it is time for a pedicure • If a dog's toenails touch the ground, they are too long • Excessive grooming is a sign that a cat or dog is meticulous

15. Aging Pets   219

You can't teach an old dog new tricks • Aging pets do not need exercise, and preventative health care (such as vaccines or heartworm tablets) no longer matters • Older pets do not develop behavior problems • Older pets do not become senile • A year in a dog's or cat's life is equal to seven human years • Pets only live about ten years • A pet that has gone blind or deaf cannot adapt and should be euthanized • A dog or cat that fails to respond must be deaf • Aging cats do not suffer from arthritis • Increased thirst and more frequent urination are normal in older pets • Incontinence (urine and/or stool) is normal in aging pets

16. Pet Death   227

The best death your pet can have is to die at home without interference • Pets suffer when they are euthanized • Mourning the loss of a cherished pet is not the same as grieving for a person • Pet owners who do not cry when their pets die do not care • When your pet becomes ill or dies, your friends, coworkers and family will understand your loss • Children do not understand when a pet becomes ill or dies • To explain the death of a pet to a young child, parents should use the terms "put to sleep" or "went to heaven" • You must be present at the time of your pet's euthanasia • You should get another pet immediately to replace the one you lost • Veterinarians perform experiments on dead pets • Veterinarians are unaffected by performing euthanasia because they do it all the time

Index   237

Author's Preface

I began my veterinary medical career in small animal (mostly dogs and cats) practice. Along the way, I acquired specialty training in animal behavior and served additionally as a consultant in pet behavior problems. I entered the honorable profession of veterinary medicine because I love animals. I like most people, too, although pets are so much easier to love. They remind us of the basic priorities and simple truths of life—spending time with those you love; patience, kindness and a gentle back scratch will bring out the best in almost anyone; sometimes you've just got to have a good howl; there are few things better than good food, fresh water and a cozy nap; unspoken understanding is far more intimate and complete than words. I remain humbled by their dignity and honored by the grace of their presence in our lives.

I have given vaccines, cardiopulmonary rescussitation and many hugs. I have watched pets lose their baby teeth, their hair and battles with terminal illness. I have listened to heart murmurs, growling pets and frustrated owners. I have held newborn kittens and puppies, dying cats and dogs and many hands. I have discovered behavior problems that were thought to be medical disorders and uncovered medical problems that were dismissed as misbehavior. I have seen aggression between pets, pet aggression toward people, and human aggression toward pets. I have known pets that had problems with house training, others that were destructive in the house, but many more that made their house into a home. And through it all, I have been inspired by my patients and even some of their owners. On a daily basis, I wish that pets could speak.

In the privacy of each professional encounter during general and specialty practice, I have been touched and angered, tickled and saddened, delighted and appalled with some of the comments and questions from pet owners. At every opportunity, I have attempted to dispel misbeliefs, misconceptions and misinterpretations regarding pet care. My collection of these common concerns has grown to fill this volume and more. And I hear new ones every day! This book is meant to entertain, inform and provoke the reader. It is not intended to be read through from cover to cover in one sitting, although you are welcome to do so. It is meant to provide concise and clear responses to some of your misperceptions without requiring you to read pages and pages of tedious theory and confusing prattle. In the busy lives we all lead, my hope is that this will then leave you with more time to enjoy with your pet!

Stefanie Schwartz, BSc, DVM, Msc
Veterinarian, Pet Behavior Consultant
Brookline, Massachusetts

1

General Misbeliefs About Cats and Dogs

Dogs bond more strongly to their owners, and cats bond more strongly to their home territory.

No one can tell what a cat or dog really feels. All we can do is make educated guesses from a human point of view of their behavior.

It has been said that cats adjust better to separation from their owners than do dogs, but this has never been demonstrated in any comparative study. Much of a cat's facial expressions and body postures are very subtle. It is not possible to judge whether a cat feels separation anxiety more deeply compared to a dog, and the notion that cats are more attached to territory than to people is unfounded. Cats are more likely to be left at home alone, with adequate provisions of course, for up to a couple of days. Both species often tolerate separation from their owners better if allowed to remain in familiar environments at least in part because they need not adapt to as many changes as they would in a boarding kennel, for instance.

The way an owner raises and continues to interact with any given cat or dog will bring out a pet's inborn predisposition to behave in certain ways. Some characteristics are less changeable regardless of an owner's efforts. The intensity and manner in which a pet displays affection and individual preference (or avoidance) for certain types of handling often are noticeable even in the first few months of life.

When characterizing what people call emotionality or demonstrativeness, some cats seem more like dogs compared to other cats and even compared to some dogs! Dogs can be stoic, independent and more interested in patrolling the yard than playing a game of catch. The alleged difference in emotion between cats and dogs is usually offered by people who are inexperienced or misinformed.

1

There is no question that cats are not dogs, and dogs will never be cats. However, to say that one species or the other is a more affectionate or attentive companion is to overlook differences between individual pets as well as between individual owners. In some cases, the stereotypes of the independent and uninvolved cat, and the dependent and obedient dog, are upheld. These, however, may well be the exceptions. Most individuals in both species range widely between these two extremes.

Animals don't dream.

Many animals other than human beings are probably capable of dreaming. The question of conscious thought and behavior with intent to accomplish a goal in other animals is still controversial because it remains difficult to prove scientifically. The ability to interact with the environment as well as learn from and remember daily experiences appropriate to their life functions is shared with other life forms. Given the many parallels, including the anatomical similarities of the brain and patterns of behavior between animals and people, it is likely that thinking exists in other species.

During the phase of deep sleep referred to as REM (rapid eye movement), people report dreaming experience. In similar sleep stages and brain activity, other animals show parallel experience. Puppies, for instance, are

We know that other animals dream during sleep, but we can only guess what they dream.

frequently seen to twitch and move paws and legs in movements that clearly resemble running. They may whimper, growl, even bark. This experience does not always seem to be pleasant, and they may wake with a start. It is not possible to know exactly of what they dream, but dream they do. Cats and kittens also show phases of activated sleep. Twitching whiskers, flicking tails and sucking motions of the mouth are not uncommon. Mature cats and dogs seem to experience less active dreams, but deep sleep associated with motor activity persists throughout their lives. We can only hope that most of their dreams are happy ones.

Dogs or cats with blue eyes are always deaf.

Individuals that have blue eyes and also have a white coat color can be hearing impaired. The genes responsible for total deafness or partial deafness (over a restricted range of sound) are frequently associated with those genes that determine the eye (blue) and coat (white) color in cats and dogs.

Congenital deafness in dogs or cats can progress over time or remain constant and may affect a limited or more extensive range of sound. A parallel phenomenon in people is called Wartenburg's syndrome. People with a white streak of hair at the hairline near the face frequently carry a gene that causes congenital deafness.

Many cats and dogs with blue eyes, however, hear and see very well. A pet with just one blue eye may be deaf only in one ear, or suffer little or no hearing loss. This individual may be completely unaffected but may carry the gene that could be passed on to affect hearing in any offspring.

Deafness can also be caused by viral and bacterial diseases that affect the ear or the brain areas associated with hearing. Exposure to certain toxins or medications can affect hearing in a pregnant mother or very young animal. Old age and exposure to very loud noise is frequently associated with hearing loss. Some cat and dog breeds with a variety of eye and coat colors have been reported to have family predisposition to congenital deafness. Dog breeds with a reported genetic basis for deafness include those with merle coats such as the Shetland Sheepdog, Collie and Australian Shepherd. Hearing-impaired Dalmatians, with brown or blue eyes, are still not uncommon.

Deafness is difficult to assess in animals without the specialized equipment so readily available for use in people. Many of the major veterinary teaching hospitals offer hearing tests for suspected problems in pets. Deafness can also be diagnosed indirectly by neurological deficits or the discovery of an occluded ear canal, for example. In many cases, profound unilateral or bilateral deafness can be suspected in a pet that vocalizes excessively or in a somewhat unusual voice. Hearing impairment may be first suspected in young pets that seem disobedient and difficult to control (although disobedience and excitability certainly can occur in otherwise normal pets).

If you are concerned about your pet's hearing, arrange an appointment with your veterinarian. Even if your pet is deaf, the impairment may serve to motivate your vigilance under specific circumstances. In every other way, a hearing-impaired animal will compensate with other healthy senses and still be a loving companion.

Cats and dogs see in black and white only.

Actually, cats and dogs do have the ability to see the world in color. The anatomy of both the canine and feline eyes suggests that these animals possess a limited color vision. It is difficult, however, to precisely correlate neuroanatomy with an individual's perception of color because seeing in color is strongly influenced by the conscious interpretation of what is being seen.

Cats may possess a greater color vision spectrum than dogs. This is based on the finding that cats have a higher concentration of specialized receptors in the retina, much like the retinal structure in people. Although cats probably do see the color red as we do, our understanding of their vision can only be inferred from their visual anatomy. Since our pets cannot tell us how they interpret the information perceived by their nervous systems, we cannot really know what they see. In dim light, it is likely that color vision fades to "black and white" for us and for our pets. There is no doubt, however, that cats and dogs see better in the dark than we can.

To keep your pet's nose black (not pink), feed it from a ceramic bowl.

The color, or pigment, of a dog's or cat's nose is determined primarily by heredity. Nose pigment may or may not be uniform. Some pets have all black noses, and others can have rosy shades with freckles. Albinos typically have no pigmentation whatsoever, and their noses will be truly pink. Generally, dogs and cats with darker coats will have dark noses and those with lighter coats will have lighter pigment. There are notable exceptions. The Samoyed, for example, is a white-coated nordic dog with dark pigment around the eyelids and on the lips and nose.

The pigment on the nose can remain constant through the pet's lifetime, or it can change with age. Sometimes, a pup's black nose can turn a rosy shade

Nose pigment is a genetic trait that is a function of breed and lineage.

as an adult. Pigment can also change because of disease. Pigment changes seen as discolored patches, for example, can be due to superficial cuts or scrapes or insect bites. Allergy, especially triggered by contact irritants, can also affect the animal's nose color. The only time that a pet's dish might impact on nasal pigment is if the individual is sensitive to plastic food bowls. Stainless steel food and water bowls may be substituted for these individuals.

Immune-mediated diseases and certain forms of cancer can affect skin pigment. For this reason, pigment change in a cat's or dog's skin should be reported to your veterinarian.

A dog's or cat's name can never be changed.

Dogs and cats learn to associate a word, which we call a name, as a signal that we want their attention. By rewarding a dog's or cat's response to a name with a caress, a food treat or a game of "catch," the pet is given positive reinforcement for responding to an assigned name.

A pet's name is a learned sound cue which, therefore, can be unlearned and relearned as necessary. This might happen when it is adopted by a new owner. Sometimes a pet's name becomes undesirable. Occasionally, a pet is given a gender-specific name ("Linda," for example), and later it is discovered to be of the opposite sex. If a pet is found as a stray, the original name may never be known by the new owners. Regardless of why a pet's name is to be changed, there is usually no problem in doing so. Even older adult dogs and cats can learn to respond to a new name.

The best way to change a pet's name is to link both original (e.g., "Linda") and new (e.g., "Max") names together. When you call the pet to play, or go for a walk, or for a meal, say "Linda-Max." After several weeks, switch the order of the paired names, calling "Max-Linda" for an additional two or three weeks. This transitional step may not even be necessary. Progress is often rapid, and many pets respond to their new name alone within days. Even if the original name is unknown, a new name can be used to attract the pet by associating the name with a positive outcome. Choose a name, and be sure to say it in a happy and light tone of voice to get attention. When the pet responds by looking at you or approaching, give lots of verbal praise (Good dog!) and, perhaps, even a small food reward. Through repetition and reward for responding to the new name, a pet usually adjusts quickly.

You can predict how big a puppy or kitten will grow by paw size.

Although the size of a young animal's feet can be helpful in predicting adult size, relying on foot size alone can often be misleading and unreliable.

It is important to consider the animal's breed, body type (which is often a consequence of breed), how large the parents were, as well as health and nutrition. A Bernese Mountain Dog runt may be small compared to littermates, but with proper care and time could well exceed their growth. A Bullmastiff pup infested with internal parasites may not grow to full potential if the worms are not eliminated.

Predicting adult size is even more challenging in cats because there is less variation in size between breeds. A notable exception is the Maine Coon which is noted for larger size, however, many domestic cats can become significantly larger than some Maine Coons.

Other factors, such as age and gender, must also be considered. Males tend to be larger and more heavily built than females even when their paw sizes compare at the same age. Dog breeds such as the Newfoundland and Basset Hound, produce stocky pups, but a Newfie pup will soon outgrow a Basset pup despite a similar paw size. On the other hand, a Basset Hound pup may have enormous feet but will not reach the adult height of a Keeshond of the same age with smaller feet.

Nature can also play tricks with a pup's predicted potential body size. A German Shepherd Dog pup with huge paws and massive parents may appear normal until two or three months of age, but if it is affected by the gene for dwarfism, it will not develop like normal siblings. Injury such as compression fractures to the growth plate of a pup's leg, for instance, may

Goliath, a Chow crossbreed pup, grew into his name.

stunt growth of the affected bone. Some tumors and genetic disease in young animals will lead to delayed closure or premature closure of the growth plate in the long bones of the legs. This can result in dwarfism and associated medical problems. On the other hand, some of the most popular breeds have been created by using inherited growth disorders, including the Bulldog, Dachshund, Basset Hound and Pekingese.

Cats and dogs have short attention spans.

Pets, like people, are individuals. The concept of attention is an interesting and controversial one. In pets, it can be difficult to assess whether or not a pet is paying attention to the owner or something else. A pet can appear to be inattentive and looking away, yet still maintain an awareness using other senses.

Any discussion of attention is incomplete without a mention of just what is attractive to an individual. Selective attention implies that one dog might reluctantly perform Obedience skills on voice command alone, but will be a whiz if motivated by a spoonful of peanut butter or some other favorite snack. Similarly, a cat might not respond to being called until hearing the sound of the can opener. Thus, a cat's or dog's attention will be influenced by past experience, certain stimuli, distractions in the environment and owners' determination to gain and maintain their pets' attention.

Hershey (Hershel Walker), a chocolate tabby with white markings, in a rare moment of repose.

2

General Misbeliefs About Cats

Cats have nine lives.

Although the divinity of the cat has been proclaimed for centuries, the cat has but one life. This myth is really a comment on the cat's extraordinary agility and athleticism which often allows repeated escapes from precarious situations. Owners not believing their pet's mortality should think twice about letting their cats outside, exposing them unnecessarily to the dangers of city streets.

Dangers exist at home, and while some are obvious, others are less so. Make sure that window screens are secure so that cats cannot fall from or push through them. Do not assume that your pet would never fall (or jump) from a balcony. Be vigilant, especially with curious juveniles, when it nears a candle flame or hot stove. Young cats can sometimes find themselves closed in the refrigerator or even a hole in the wall. Parents should supervise children and teach them how to gently and appropriately interact with pets so that a pet's tolerance limit is not tested.

The cat is endowed with superb vision and depth perception as well as acute senses of hearing and smell. Additional environmental cues are provided by whiskers sensitive to vibration and air currents. Despite their keen senses and muscular coordination, cats are vulnerable creatures whose talents are occasionally sorely pressed by life's challenges. We need not expose them to avoidable risks regardless of how well they may be able to surpass our realistic or idealistic expectations.

Cats are not demonstrative pets.

Surely this notion can exist only in the minds of people who have never owned a cat. Even if they are not "lap cats," most cats will at least remain near the owner and solicit affection on their own terms. Occasionally, cat owners complain that their pet seems to be too clinging, shadowing the owner's every activity and seeking close contact whenever possible. Some cats retain a more wild than domesticated nature and resent or avoid interacting with people. But then, there are some dogs that behave in similar fashion. There are those who maintain that nothing compares with a good dog. A good cat knows better!

Cats purr only when they are happy.

Purring is an indication of a cat's contentment. Queens, for example, often purr while nursing their kittens. Few things are more gratifying to cat owners than to listen to a cat purring. It is somehow an honor to pet a cat, or to engage in some other pleasurable interaction such as grooming, and to be rewarded with these pleasing vocalizations.

But cats will also purr when they are anxious or even ill. In these cases, purring may be an anxiety-releasing mechanism that may help to comfort the individual. No one really knows how purring is produced although it certainly emanates in the back of the throat. We may not fully understand how or why cats purr, but we are very glad they do.

Cats wag their tails only when they are angry.

The tail can shed much light on a cat's state of mind and current motivation. In general, the grander the movement the more aroused the cat.

A cat's tail signals a wide range of emotional states. The tip alone or the entire tail can move provocatively or with indignant agitation. It may be held high and quiver in anticipation of friendly attention or just prior to urine-spraying. The tail may be held high with hair standing on end or may be tightly tucked beneath a fearful animal's crouched body. A cat may gently wave the tail from side to side while drinking or eating. Cats may also lay on their sides in a stream of sunlight and intermittently wave only the tip.

The movement and posture of the feline (and canine) tail should be interpreted along with the individual's facial expression and body posture as well as the context of the situation. Rapid, wide and arrhythmic twitching of a cat's tail indicate nonspecific arousal. When displayed in a growling cat undergoing a veterinary examination, for instance, this may warn of impending aggression. A similar tail motion is also seen, however, when playful cats ambush or stalk one another. The tail also signals the individual's psychological and physical well-being. Held erect, it is like a flag of confidence in the socially assertive cat. The same animal may carry it low, half-mast if you will,

Affectionate greet-ings between cats are often very subtle and brief, but no less significant than more obvious displays typ-ical of other species.

when it is anxious or not feeling well. Given the wealth of information transmitted by a cat's tail, it might make you wish we still had tails, too, doesn't it?

Cats need catnip as part of a healthy diet.

The plant commonly known as catnip is a member of the mint family. When dried, its leaves are easily crumbled and have a faint smell not unlike alfalfa hay.

There is nothing in catnip that is required for balanced nutrition or the emotional well-being of felines. Cats do not need contact with catnip in any way. It is our own intrigue with the cat's reaction to catnip, along with clever product marketing, that makes it so popular.

Catnip contains the compound nepatalactone, a natural hallucinogen, that is chemically related to the active ingredient in marijuana. When a cat inves-tigates catnip, molecules of nepatalactone are inhaled. Although its precise action on the central nervous system is not clear, the response to catnip is well described. During the catnip response, the cat seems to roll around in apparent delight (or what we interpret to be something akin to it). There is no recognized danger in exposing a cat to this natural psychoactive substance, and it is rapidly eliminated from their systems. Still, until more is understood about it, exposure to catnip should probably be moderated.

All cats love catnip.

The catnip plant (*Nepata cataria*) grows wild in North and South America. All members of the cat family are reported to be sensitive to the plant's natural hallucinogens. This is interesting particularly because although the plant does not grow on the African continent, the great cats of Africa are catnip-sensitive. It was once thought that only certain cats reacted to the catnip plant, but more current information shows that all cats will exhibit the catnip response if conditions are right. Kittens have no catnip sensitivity prior to reaching puberty. Cats that are anxious or ill or distracted may show an abbreviated reaction or, more likely, none at all. It is seen in males and females and is unaffected by neutering. Not all cats seem to enjoy the catnip experience. Some individuals actively avoid contact with it despite an interest and sensitivity to catnip in the past. A strong voluntary component is implied.

The catnip response lasts an average of six minutes. The pet can usually be interrupted from this apparently pleasurable experience and does not seem disoriented. After exposure, the cat can be seen to sniff (sometimes sneeze), lick, chew, roll and rub itself in the scent. The cat may also scratch or dig at the odor's source, and it is this element of the sequence that has been applied to facilitate scratching-post training. Dried catnip can be directly applied to scratching surfaces or catnip-treated toys can be suspended nearby. Following the peak response, cats will lie down in apparent contentment. Other cats launch themselves into another form of play prior to resting. Although the catnip response may be remotely suggestive of sexual pleasure, it may be more consistent with behaviors related to hunting and predatory play.

Cats are always good hunters.

Members of the cat family rank, as a group, among the most superbly adapted predators. They possess exquisite eyes, not only in esthetic beauty, but also in function.

Among mammals, the feline visual system has no equals. Their eyes are supremely adapted to detect motion even in dim light. Erect and mobile ears directed to sound provide the best reception for sharp hearing. Keen olfactory sense and the detection of minute vibration facilitate the hunter's advantage in seeking prey.

The cat is also equipped with a full arsenal to fell victims. A sinewy sprinter with agile and taut muscles give speed and power. The teeth are those of the ultimate carnivore, combining those intended to puncture with others intended to gnaw or shred. Flexible claws spike and lacerate. The domestic cat shares all this with untamed relatives *but may not have inherited the instinct to apply them.*

The cat's hunting ability is a learned and practiced talent based on inborn instinct. Genetic and learned components of hunting together create a mastery that is unique to each animal. Not all domestic cats are equal as hunters, and some are not inclined to hunt at all even when given the opportunity. This range between pet cats is due to domestication which, by definition, implies human interference in the need for survival mechanisms. Pet cats do not need to be good hunters in order to eat or to provide for offspring. Selective breeding by people has focused primarily on their external appeal. Hunting prowess, therefore, is of secondary survival value and so may or may not be inherited. Individuals that inherit both the talent and desire to hunt may be weaned and separated from their mother. This restricts the important contribution of maternal instruction in the field. Other cats may have every opportunity to roam outdoors and fulfill the promise of their predatory ancestors but have limited genetic predisposition. A cat might seek and capture prey but not kill it, or it may make the kill but not consume it.

Cats must roam outdoors to live happy lives. (It is cruel to confine a cat indoors.)

It can take a lifetime for us to realize happiness and fulfill our needs and dreams. How can we judge what will be emotionally gratifying for another human being let alone another animal? We can only guess what will make our pets "happy," through human eyes and human values, and based on their behavior. The only thing of which we can truly be certain is the reality and multitude of dangers beyond the relative security of home.

The domestic cat is a descendant of small wildcats which still roam in the forests of Europe and African brush. Their predatory nature and untamed spirit lives on in many domestic cats. The urban jungle is not Nature's jungle. Concrete, traffic, dense human and cat population on every city block, and concentrated epidemics of contagious illness in no way resemble the life cats were intended to lead. Indeed, these elements do not present a safe environment for the human condition either. Why is it cruel to protect these innocent and vulnerable animals from dangers that they don't perceive or comprehend? Isn't it more cruel to intentionally expose such a creature to these dangers?

Every intelligent mammal seeks to expand territory and stimulate intellectual curiosity. The instinctive desire to roam outdoors varies in domesticated cats and is determined, in part, by the opportunity to investigate the outside world and the consequences of this effort. If access to the outside world is blocked, most cats will adjust to the space inside your home, and will live a safer, healthier life. They will not contribute to the tragic pet population explosion. They will not cost their owners avoidable veterinary fees due to injury or infection. They will amuse themselves in your home with alternative activities and objects of amusement. Their "happiness" will, in part, be a function of yours.

House cats find alternative distractions like this fish tank, a "feline entertainment center."

An outdoor cat can never adapt to confinement as a house cat.

The cat is a highly adaptable creature. The key to any being's survival is the ability to cope with change which can sometimes be sudden and even lethal. In contrast, merely restricting a cat's territory to the boundaries of your home is not a life-threatening gesture.

The decision to keep a pet indoors is a gesture of kindness and responsibility. The cat should grow accustomed to this, if given the time. Any change is stressful, regardless of whether the change is perceived as positive or negative. The key to making the transition is in remaining steadfast. Some cats, particularly those who prefer to remain confined during cold winter months or rainy days, may not be as difficult to convince to stay inside once the decision is made for them. Others must be supervised closely and prevented from making escapes. It is most important that escape attempts be unsuccessful because once a cat knows escape is possible, attempts will persist for a while.

Although every case is different, there are basically two ways to convert your outdoor cat to a house pet. The gradual method calls for restricting access to the outside by initially delaying the time, just by 15 minutes more each day, the cat is let outside. Be sure the cat does not learn that vocalizing long enough or loudly enough forces you to comply. Try keeping the cat confined for one day a week. The next step would be confinement for two days each week, one day at the beginning and another day later on in the week. Eventually, add a third day of confinement. This process should take many weeks or even several months. Make the transitional period as smooth as possible.

The other method is more drastic and requires your pet to go "cold turkey." For the determined outdoor cat, this method will be the most stressful and anxiety may run high. For this cat, ask your vet about anti-anxiety medication.

During the adjustment period, it is important to remind yourself that your cat will be far better off in the long run. Even if your pet seems restless and preoccupied, refrain from letting your guilty feelings overwhelm you. Instead, provide your cat with alternative and attractive activities:

- Divide daily portions of food and leave smaller portions in new places around your home.

- Spend extra time playing with or brushing your pet especially before your bedtime.

- Purchase or make new toys.

- Start an aquarium of fish. Although the tank top should be secured to prevent fishing, this "feline entertainment center" can keep your cat's interest as well as add to your home decor!

With your patience and persistence, your cat will become a homebody. It will be safe from contagious disease, cat fights, dog fights, cars, trucks, buses and people who might do them harm. Your cat can and will adapt to remaining indoors. The rest is up to you.

Cats always land on their feet.

Feline agility and sensorimotor coordination are nothing less than phenomenal. Cats have a marvelous ability to right themselves, or direct themselves, toward the ground in preparation for landing. A sense of balance along with a flexible spine and muscle tone combine to give great advantage when falling from heights.

However, the cat is far from immortal. A cat can be severely injured or killed even falling from a first- or second-story window or less. This may be because they do not have enough time to assume the ideal posture to prepare landing to prevent injury. In contrast, cats may have relatively few injuries sustained from a fall many stories high. The "high-rise syndrome" refers to cats that fall (or perhaps jump) from high-rise buildings. It has been suggested that the reason fewer cats are not killed during these incidents is that they have time to anticipate the ground during their flight. By spreading out their four legs, they may create aerodynamic lift and, consequently, slow the plummet.

Cats see better at night than during the day.

Cats' night vision is not better than their day vision, but they definitely see better at night than we do. The cat possesses the largest eye of any carnivore, which implies both reliance on and acuity of vision.

The eyeball is essentially a fluid-filled sphere. Light passes through the pupil and the lens of the eye and is then focused on the rear inside lining of this sphere. This lining at the back of the eye is "carpeted" with a special layer of receptor cells called the retina. Retinal cells collect the reflected light waves and transfer them into electrical impulses to be transmitted through the optic nerve of each eye into the brain.

The cat's retina has a high concentration of receptor cells called "rods" which are particularly effective in very low illumination. Beneath these layers, part of the tissue is pigmented, appearing as a beautifully colored mosaic, which helps to further absorb light and extract even more sensory detail in dimly lit settings. This area is what makes cats' eyes seem to glow a greenish hue in dim light. These adaptations evolved to enable the cat to hunt during the day and at night. Some cats, such as the Siamese, lack the pigment in this layer at the back of the eye and this reflects a reddish color due to blood vessels in the eye that are hidden when pigment is present. Many dogs also possess a pigmented retina and many rod receptors. However, these do not compare with the exquisite visual anatomy found in cats.

Trimming a cat's whiskers will cause blindness.

Cats are blessed with acute stereoscopic vision, a relatively wide field of view, and limited color reception. At very close range (e.g., when a toy or prey are reached) its ability to focus may not be so refined. It is at this point that the cat must rely on other senses. Hearing, smelling and vibrations detected by the whiskers become essential to compensate for cats' shortsightedness.

The cat's whiskers are highly specialized facial hairs that sense air currents and disturbances from vibration. Whiskers have small muscles at the base to point toward an object or point of interest. Each whisker sends incoming information to its own specific area of the brain, where this data is rapidly processed and determines the individual's next move. Trimming a cat's whiskers will certainly be uncomfortable, and it will unnecessarily deface these long and lovely accessories.

Abundant hair in the ears indicates the cat is male.

Many female cats (and dogs, for that matter) have hair growing inside the ear and even on the outer portion of the ear canal. Hirsutism, the degree of hairiness, is not a secondary sexual characteristic in pets as it is in physically mature men. The notable exception is the mane of hair that frames the male's face in many cat species. The most obvious example is the lion's mane, but the male tiger has a mane, too. Tomcats, intact or neutered after puberty, develop a more pronounced hair growth with thicker skin on the head as well as a sturdier physique compared to the female. Hairy ears is a quality that is related to the breed (Persian versus Siamese) or to the individual (long-haired or shorter coat). Of course, the best way to determine a cat's sex is not to look at the head at all. The opposite end will generally provide more reliable clues!

Stripes on the tail indicate a cat is male.

Nope. Many female cats have striped markings on their tails. The only coat color that is specific to sex is the calico cat, which comes in either the traditional tricolor (white, orange, black) or blue cream (gray, white, cream) varieties. The calico cat is almost always (ninety-nine percent) female. The rare male that is born with a calico coat will be sterile.

The domestic cat, correctly described as a tabby cat when having striped markings, is descended from a small wildcat and is only distantly related to tigers and the other Great Cats.

Striped markings indicate a cat is descended from tigers.

The black-striped tiger is a large and persecuted member of the cat family. It is composed of several varieties in ever-dwindling numbers in Asia, India and Siberia. The domestic cat is a distant relative of the big cats but is a more direct descendant of much smaller wild ancestors. The correct term to describe domestic cats with striped, swirled or spotted markings is "tabby" and not tiger.

    Although the exact origin of the domestic cat is unclear, it is most likely related to the diminutive African wildcat or European wildcat, or both. These wild tabby-marked cats still live today and occasional matings between them and their domestic cousins continue to be fertile. This suggests that, genetically at least, they are not that far removed from each other.

Black cats are aggressive and unlucky.

Although there is some suggestion that the black coat color in mink and foxes is linked with a somewhat tamer temperament, this evidence does not exist for dogs or cats. Black cats may not appeal to pet owners for many personal reasons, but superstition is lost in those of us who admire the glossy sheen of a black cat.

During the Middle Ages in Europe and America, cats, and black cats in particular, were considered to be demonic embodiments and were associated with satanic practice and witchcraft. Cats frequently suffered the same torture and torching as people, predominantly women, accused of practicing the "black arts."

Today, crossing paths with a black cat still makes some people cringe. Medieval superstition is alive and well in many forms (and sometimes perpetuated by modern "witches"!). Luck, whether fortunate or misfortunate, is a force that seems to affect events, objects or outcomes in a person's life. Luck is, in part, what you make it. If you have a friendly and healthy black cat you are, indeed, lucky.

Your backyard rapidly becomes a lonely, negative and sterile environment for your dog.

3

General Misbeliefs About Dogs

Dogs are direct descendants of wolves.

The earliest evidence of the coexistence of dogs and people can be traced to early settlements at least 10,000 years ago. It is unclear whether the first individuals to be domesticated were wolves or a smaller relative of the wolf. Some scholars suggest that other wild members of the dog family, such as jackals or their ancestors, contributed to the widely diverse breeds of domestic dogs that we know and love.

It can be difficult to recognize the common ancestry in a Chinese Crested dog or the Chinese Shar-Pei, yet it lives in the spirit of these breeds as well as in the German Shepherd Dog and the Borzoi. Despite this tremendous physical divergence, the fundamental behavior of dogs is generally consistent among breeds.

The social behavior of the wolf and wolf pack provide great insight into understanding our pet dogs. Much of a dog's social behavior, including its interactions with people, echo the dynamics of the ancestral pack.

Dogs do not need to be walked on a leash if they have access to a fenced yard.

There are few things more jubilant than a dog's celebration of an invitation to go out for a walk. Just say the word (some dogs even learn the connection if it is spelled out instead "w-a-l-k"!), and you will see joyful prancing. Eyes brighten eagerly and seem to say, "Gee, do you really mean it? Hurry up, let's go!" Even if a dog is lucky enough to have access to a yard, the need for daily walks in the company of a human, usually the owner, remains constant. Regardless of your schedule or energy level, your daily commitment continues for the lifetime of the pet. Pets cannot understand that a stressful day at work left you with the sole desire to crawl into a hole somewhere and hide from the world. They wait patiently for you to come home all day! Taking your dog for a walk is the best remedy for both of you after a long day.

There are many important needs answered by walking your dog. Walking your dog immediately following each meal is essential to maintain desirable house training. This gives the dog reliable opportunities to void in desired locations and gives you the opportunity to praise appropriate behavior. Walking your dog is obviously good exercise, but it is also intellectual stimulation. There is only so much interest an intelligent animal can derive confined within the walls of your home and even your yard. During a walk, your dog can investigate all the smells, sights and sounds of your neighborhood, process these environmental details and interpret any changes. Given the acuity and range of the canine senses, dogs experience the world around them to a much richer degree than we do, if given the chance.

From the dog's point of view, a walk is a cooperative form of territorial patrol and strengthens the bond between you to promote "pack" unity. It also gives you the time to unwind in the company of a good friend. During walks, Obedience skills can be practiced under distracting conditions to ensure reliable training in case it is urgently required. This, in turn, complements the intellectual "exercise" afforded by each outing. So, what are you waiting for? Where did you put that leash?

A dog can be left alone in a fenced yard.

For those dog owners who are also home owners, a yard is an additional blessing. It is important, however, to continue to walk your dog on a daily basis. You and your dog can enjoy the yard and the benefit of private space and open skies overhead. You can use this area to practice Obedience skills under more distracting conditions, thus challenging the reliability of your pet's skills. You can play with your dog without fear of breaking a lamp or sending crystal crashing. You can oversee your domain as you sip your morning coffee or afternoon iced tea with your favorite canine companion. But your dog still needs to be walked.

Enjoy your pet and your yard but be vigilant. Never take safety for granted by assuming that the fence is a magical guarantee. Don't take for granted that your yard is a protected haven. Fences are artificial barriers that may succeed in keeping your dog in, but hazards can always breach the barrier of your fence. Most fences will not deter another person or animal intent on gaining entry. For instance, there are few things more motivated than a male dog keen on reaching a bitch in heat. Burglars have both experience and desire to cross over most obstacles. A dog is often more vulnerable than we think against the threat of an intruding and aggressive dog or person. If you are confident that your dog will reliably defend self and property, then you may be relatively unconcerned about these risks. On the other hand, you will be responsible for any injury to your unsupervised pet, and you will also be liable for harm that befalls trespassers, such as children who may climb over or under a fence to retrieve a ball.

It is tempting to send your dog outside in the yard when you are busy, but this is no different from parents who rely on television to occupy children. A yard is simply not enough for your dog and could even become a type of social banishment. Preoccupied or complacent with a false sense of security, you could easily forget your dog is in the yard until you hear barking, or worse, realize you have not heard any barking for a while. Although fences may successfully keep problems out of your yard, the problems inside your yard will continue. An active and intelligent animal needs much more physical and intellectual stimulation than that which is available in your backyard, no matter how large the area. Given enough unsupervised time, your pet could direct energy to digging holes, uprooting your flower garden, and discovering or creating a weak point in your fence. Once your pet succeeds in leaving your property, most pets will not remain nearby. A dog will make repeated escape attempts once achieving success. Much of what motivates and entertains a dog stems from social interaction. Left alone, the dog could remain relatively inactive (which is not healthy) or seek undesirable activities.

It is okay to leave a dog tethered in your yard.

"Runners" can be attached overhead or to a point near or at ground level. Tying your dog to a runner (long lead, rope or chain) will restrict movement within the yard and could prevent the dog from digging or running away.

There are, however, some serious considerations. It is critical to check periodically whenever a dog is left outside and unsupervised, even if tethered.

- The dog can become tangled around a tree or some other obstacle and remain helpless.

- A tethered dog may not be able to reach shelter or escape unfriendly intruders.

- You or other friendly visitors in your yard can be injured if you trip across the line.

- The "runner" can be particularly hazardous when anchored to an elevated point. Overhead "runners" can be wrapped around someone's neck, especially children, and your unfortunate pet could even strangle.

If you feel comfortable with the risks of tying your dog to a "runner," at least make sure that it is secured at ground level and check frequently to make certain that all is well.

"Unseen fences" (that deliver painful shocks or high-frequency sound) are foolproof.

For most pet owners, the purpose of a barrier is to reliably restrict a dog within property boundaries. Some pet owners are not satisfied with a functional fence and also require one that is inconspicuous and does not interfere with the view or appeal of their home. Concealed fencing sounds like the ideal solution.

The problem is, however, that these costly systems fail to contain many pets, and it is difficult to know in advance if these expensive barriers will be successful. Some dogs comply with the invisible barriers, but most dogs are likely to ignore the "invisible" boundary. Although some pets are initially contained they later discover that the brief punishment incurred by crossing the hidden line is worth the risk. Concealed fences work better for small dogs or those with less assertive temperaments or dogs that are just not very determined to escape the imposed confinement.

Two systems that rely on transmitter collars around your dog's neck are currently available. One of these delivers painful electric shocks of low, medium or high intensity. The other works by the emission of painful levels of sound that are inaudible to most people. If you are considering either of these systems, it would be worthwhile to visit someone who has already installed one on their property. Ask if you could test it on your dog, that is, if you feel comfortable with intentionally allowing your pet to suffer an electric shock or painful auditory stimulus. If there is strong motivation, your dog will cross the hidden boundary. If your dog even accidentally triggers the punishing mechanism it could discover that the punishment is bearable and does not prevent committing the crime. Once your dog learns that the barrier can be crossed, increasing the intensity of the "punishment" may not provide further restraint.

There is a greater chance of failure than success with these systems. There is the additional risk of making your dog neurotic or fearful. If your pet triggers the punishing shock when running to greet you, for example, the dog will learn to avoid approaching you. If the shock is delivered at the moment a bird calls, your dog could become phobic to birds. Your dog could also learn to resent anyone or anything touching around the neck or head. The potential injury to your pet from electric shock collars cannot be understated. Dogs have been severely burned by these instruments. Although the sound-based collars seem to be more humane, they are also likely to be less effective. Failure of these invisible barriers can result in your dog's escape and harm. The risk of physical injury to your pet as well as the significant risk of investing in these devices is yours to consider.

Purebred dogs have better temperaments than mixed breeds or "mutts."

For most pet owners, the definition of a "good" temperament implies no aggressive tendencies, particularly toward people, or low intensity aggressiveness that can be triggered only with difficulty. It is difficult to prove that any breed of dog, or cat for that matter, has a uniform temperament, good or bad. A mixed breed dog could have a more even temperament than a purebred one, and vice versa. In some breeds, aggressiveness is intentionally selected. This was the case with Doberman Pinschers years ago when the breed was popular for guarding and attack training. More recently, Rottweilers and American Pit Bull Terriers have gained notoriety thanks to the efforts (or lack of them) of misguided people.

Although genetic predisposition is fundamental, individual training along with responsible ownership are equally significant. A breed should not be condemned because some of its representatives are undesirable, neither should we condemn all the people involved in breeding the dogs. Unstable temperaments occur in many breeds as well as dogs of mixed breeds.

For the average dog, regardless of breed, each individual inherits a tendency toward a range of normal behaviors in specific circumstances. Behavior can be encouraged or discouraged by the response given or the outcome achieved. In other words, whether or not you acquire a purebred, crossbred or mongrel, you will usually get back what you put into its upbringing.

Purebred dogs are better pets than mixed breeds.

Each of us has an ideal of what makes a good pet. It could be based on our perception of canine beauty or a desire for a dog that has specific athletic ability. We might simply want a companion that is small enough to be carried. There are hundreds of dog breeds to satisfy different functions or notion of what is attractive or functional. Some dog owners like the look of a French Bulldog and others prefer the Bearded Collie. People who spend time hiking outdoors picture themselves with a Labrador Retriever while others just want to cuddle with a Pug.

The truth is that many people want a dog that somehow symbolizes their own unspoken desire to be attractive or popular or dominant. These desires can be satisfied by mixed breed dogs too. The majority of dogs care little for their own size or appearance, or for that matter, our size or appearance. Purebred or not, we all focus entirely too much on the physical aspect of our pets and ourselves. Mutts combine the physical and behavioral attributes of their parentage. Mutts or purebred dogs not of close breeding may be less prone to congenital illness and inherited temperament problems than dogs of tightly bred, concentrated bloodlines. As long as your dog is loved and cared for, a pet's physical appeal is secondary.

During car rides, dogs should be allowed to have their heads out the window for fresh air and to enjoy the drive.

A dog's head protruding from a moving vehicle is a medical accident waiting to happen. Airborne particles of dust can hit your dog at high speed. Debris, accelerating like speeding bullets, can enter your dog's eyes, ears and nasal passages. Roll the window down enough to let air in yet still keep your dog's face inside a moving vehicle. It is your job to prevent your pet's unnecessary exposure to injury.

It is safe for a dog to ride in the back of a pick-up truck.

Pick-up trucks are practical vehicles meant to transport any number of objects. Dogs, however, are not objects and should not be included in the list of cargo. The animal is at high risk of being thrown around or even out of the vehicle if the driver must swerve or drive over uneven roads. The canine passenger is also exposed to adverse weather without shelter. Flying debris and dust can cause serious harm to its eyes and other sensitive tissue. Leashing and tethering the dog to the pickup does not protect against many of these hazards. Many dogs are unnecessarily injured as a result of traveling in the back of an open vehicle. A dog deserves to be transported inside the cabin in safety and comfort. When there is no room for the animal inside, a sturdy fiberglass crate will provide adequate protection under most circumstances. If a pet carrier is not available, leave your dog in the comfort and safety of your home.

AKC registration papers certify a dog's quality.

Registration documents that originate from kennel clubs such as the American Kennel Club (AKC) certify that a dog is purebred. The certificates are based on information forwarded by dog owners and breeders and details are not routinely verified. Unfortunately, this leaves much room for people who seek to exploit the honor-based system. Additionally, with every registration certificate, the AKC states that the dog is purebred and registered, but this does not guarantee health or quality, only that the dog's parents were registered. Not all German Shepherd Dog puppies are show quality, nor is every Golden Retriever an ideal family pet.

Most dog breeders, but not all, are concerned with the quality of the puppies they produce and with the people who acquire them as pets. There are those who breed unhealthy or otherwise undesirable dogs under poor conditions, such as the puppy mill, yet they produce animals with AKC registration documents. Pet owners with little knowledge of sound breeding practices breed their dogs because they think it is nice to have puppies or want to make a few extra dollars. A dog's written declaration of purity, in many cases, is just a piece of paper.

It is unwise to base your judgment of a pup's quality solely on whether or not it comes with registration papers. Rely on your impression of the individual breeder based on their care and concern for the puppy and mother as well as the breeder's knowledge and reasons for breeding the litter. Inquire about their professional reputation with the AKC or local kennel club. Speak with a veterinarian before you adopt a pet about these and other concerns, and make an appointment for an examination soon after you make your choice.

Dogs that run away from home can find their own way back.

Many species of animals cyclically migrate over vast distances. Ancestral instincts direct an individual or group of individuals to move between feeding or breeding or resting areas on a daily or seasonal cycle. There are few who do not marvel at the splendor of a flock of Canada geese passing overhead. Whales travel hundreds and sometimes thousands of miles to feed in favorite waters and to bear their young in the few safe harbors. African herd animals such as the elephant, zebra and many species of gazelle persist in ancient routes despite restrictions and obstacles placed by local human populations. In North America, caribou herds still follow these ancestral drives. Buffalo once thundered in the millions across the continent but are now silenced by ignorance and greed.

The dog is not a migratory animal in the same sense as these and many other species. Domesticated from a member of the wolf family more than 10,000 years ago, the dog and wolf share many behavioral parallels. The size of the territory claimed by a wolf pack varies with the availability of and demand for food or water, population density as well as encroaching competition from neighboring wolves. The wolf's territory must be initially investigated and then routinely patrolled.

The domestic dog is certainly able to become familiar with its own home range, given the opportunity and provided the individual dog has inherited a full complement of ancestral instincts. There are exceptional stories of pets that have been separated from their owners or homes and find their way back. These are, however, exceptions. Dogs that habitually run away from home may get practice at familiarizing themselves with visual landmarks or olfactory cues. These escape artists may actually expand their territories at each escapade, incorporating new data that may be essential to retracing their path. They could also get lost and stay lost.

One side effect of domestication is the dilution or absence of many survival instincts. Most pet dogs have not had the opportunity to practice tracking skills. Many might not survive at all if removed from the support of their human "packs." It is more likely that a lost dog would remain in the vicinity or roam aimlessly in a random route. The portrayal of dogs in motion pictures that heroically return home after months or years of traveling hundreds of tortuous miles is more glorified than factual. A pet is more likely to panic or become exhausted, and retracing steps can become more and more a remote possibility. If your pet escapes the safety of your property, it would not be wise to rely on the animal's unassisted ability to return to you. Every effort should be made to rescue a lost dog or cat and to prevent the desire and opportunity to roam away.

Dogs run away from home because they are ungrateful, angry or dissatisfied with their food.

Dogs run away from home because they have the opportunity and the motivation to do so. Roaming is not a behavior that reflects a pet's ingratitude or malice. These are human attitudes that reflect human values, which are beneath those of any dignified dog. Roaming (escaping or running away) is common in young dogs, male or female, and is a reflection of the individual's intelligent desire to investigate the world.

In addition to the intellectual challenge of creating an escape route and executing the escape, it may also satisfy the dog's need for more physical activity and social interaction. Dogs that learn, even from a single isolated adventure, that roaming is possible, are likely to make future attempts.

You should punish your dog for running away.

Regardless of whether you finally find your lost dog after a frantic search or whether your dog returns home unassisted, never, never, never punish! If you greet your dog with an angry voice, stern facial expression and menacing gestures you will be, in effect, punishing your pet for approaching you. Your pet is unlikely to associate your displeasure with a remote act of running away minutes or hours earlier. Even if you catch your dog in the act of leaving your property, would you want to remain with someone who is punishing you? If you locate your dog during your anxious search, resist your natural urge to express your panic and frustration. Instead, give the "Come" command in a jolly tone of voice. If you must, run in the opposite direction, away from your delinquent dog and gleefully repeat the command to "Come" as if to play a game of "chase" (which is undesirable under other circumstances). Give a "Sit" and "Stay" command in a firmer tone, approaching calmly and without haste. If either or both of these commands fail, then you know you really have a lot of Obedience training to review or to learn. Be sure to bring a leash and a favorite treat as a backup. Sit on the side of the road or remain in visual contact with your pet. Eventually, the animal should willingly approach you, unless you have punished your dog for coming toward you in the past.

Of course, the best way to teach your dog not to run away is to prevent the desire and ability to escape.

- Secure your fence and identify escape routes.

- Do not leave your dog unsupervised for extended periods. Provide lots of social interaction and leash walks with you everyday.

- If your pet does manage to get away, revise scheduled activities and upgrade your fence.

Be grateful that you are lucky enough to be reunited with a lost pet and always greet it with open arms and enthusiastic welcome. After all, he or she greets you with nothing less!

Dogs have no preference for using a right or left paw.

Limited evidence suggests that the majority of dogs (this has not yet been studied in cats) prefer to use their right paws when taught to perform certain tasks. Almost sixty percent of dogs in one study appeared to be "right-pawed," and a little less than twenty percent preferred to use their left paws. A significant proportion of the "righties" were females, implying that left-pawedness may be more likely in males.

The significance of these figures is unclear but it indicates that people are not the only animals to have dominance of one brain hemisphere over the other. It is too early to make any conclusions based on a single study, but future research will be interesting.

On a hot day, your dog is better off in the car than at home.

Although it seems unkind to deny your dog the pleasure of a car ride and to deprive each of you of the other's company, leave your dog at home on hot days. In fact, leave your dog at home even if the temperature is just a bit warm.

Leaving your dog in a parked car on a warm day is like leaving it an oven. Within minutes, the temperature inside can soar far higher than the temperature outside the car. Rolling down the car windows to provide ventilation is not any more prudent because your dog could run away or be stolen. Additionally, ventilation provided by opening the windows even an inch or two cannot provide adequate circulation of air or control the heat inside. Even when available, shade does not keep the temperature controlled inside your car for more than a few seconds.

The risk of heatstroke or theft should convince you never to leave your dog confined in your car, for even a few minutes.

Most dogs, like this Greyhound, are not meant to spend long periods out-doors in the cold, but still need to play and exercise despite the weather.

Heat stroke (hyperthermia) due to over confinement in poorly ventilated areas occurs when normal physiological cooling mechanisms in the dog (cat or person, too) can no longer keep body temperature within normal range. Heat stroke can occur in your yard, too, if it is very hot and your dog cannot find or reach adequate shade or fresh water. If you return on a hot day to find your pet unable to rise or laying on its side and breathing heavily, the animal's life could be in serious jeopardy. This is an emergency situation that requires your immediate intervention. Pour cool water on the dog's head and over as much of the body as you can. If a garden hose is available, soak the coat. Wet towels can be wrapped around the dog's body. Go to the nearest veterinary facility immediately.

Dogs are meant to live outside in cold weather.

No animal can survive outside for very long without proper shelter. Extreme weather cannot be tolerated by even the healthiest and strongest over extended periods. Even Siberian Huskies and Nordic breeds with "double" coats, or those with long and thick hair, can suffer frostbite or perish from prolonged exposure to severe winter conditions.

- If your dog spends much of the day in your yard, provide a well-insulated doghouse.

- Better yet, install a "doggie door" so that it can come inside when it needs or wants to.

- If it is below freezing, avoid leaving your dog outside for long without checking (every half-hour or so) that the dog seems comfortable.

- Don't forget that water freezes in dog bowls outside when the temperature drops and your dog needs to drink water regardless of the weather.

Another reason why a dog should not be left outside for long, regardless of whether or not the weather seems severe, is that the dog is a social animal, not meant to be left in solitary confinement. Even if you rely on your pet to protect your property, the dog cannot be "working" twenty-four hours a day. Bring your dog inside to socialize with you. This will only strengthen the bond with you and help maintain a natural commitment to guard territory. Dogs are pets, even when they are working dogs. Do not abuse their physical limits or their emotional needs.

A dog's instinct is to please all humans.

Some dogs don't like people. Some dogs don't like some people. Some dogs like some people that don't deserve it. Every dog is an individual, born with a preference in social interactions with other dogs or people. Genetic characteristics are the basis for learning acquired over a lifetime.

Early learning has long-term effects because it shapes so many subsequent perceptions and responses later in life. A puppy not exposed to people during a critical phase of development (roughly between the ages of six to thirteen weeks) can be fearful and shy with people for a lifetime without a carefully tailored treatment program by a qualified veterinary behaviorist. A dog abused at any age may learn to avoid the abuser and recall the negative impression in similar contexts or with people who trigger the memory and response in some way. A dog that has limited interaction with people of a certain sex or race can display fearful aggression toward unfamiliar individuals.

Still, dogs that are raised in perfectly wonderful homes become shy or aggressive (see chapter 7). A dog can become afraid of a person or object without any prior contact. Another dog, raised in neglect and with cruelty, can remain loyal and devoted to less than satisfactory owners. A dog is not always a person's best friend. It depends on the dog. It also depends on the person.

4

Pet Selection

Selecting a pet is an emotional and spontaneous decision.

Selecting a pet should be a long-term investment that should not be done on a whim or in a moment of frivolity. A pet should be a companion for a lifetime.

In an ideal world, every dog and cat deserves to be cherished by responsible people who will nurture and attend to the pet's physical and emotional health. Unfortunately, this is not an ideal world we live in. If you are considering acquiring a pet, think about it again. When you have considered it carefully, think about it some more. Talk to friends and family that have pets.

Go to a library and study books about the variety of breeds of dogs or cats. Visit a pet shelter in your community and get to know the staff. Speak to a local veterinarian and even to a pet groomer. If you have not had pets since childhood, remember that your parents shouldered much of the effort and that your memory of the work involved may be unrealistic. Unless you are ready to devote yourself to a pet for that animal's lifetime, don't get a pet.

Do not make a Christmas gift of a kitten or pup (or a baby chick or bunny at Easter). Do not buy that "puppy in the window." No pet deserves to be given away or put to sleep because of poor planning. Pet ownership should not turn into an entanglement that you regret. Your pet will suffer most because of it.

The most important consideration when selecting a pet is how it looks.

Appearances are misleading. Selecting a pet based solely on physical attributes is like buying a car without a test drive or marrying someone that you have only seen in a photograph. Selection of a pet must always be based on the individual's temperament and health. Preliminary considerations of a breed's physical characteristics, adult size and your gender preference are necessary. The amount of care is a practical criterion determined in part by the dog's adult size and coat. But these are never more important than temperament and health. You will forget how beautiful your pet is if your dog or cat bites or becomes chronically ill. Physical appeal is important, but it is only the outside of the package. What really counts, after all, is what is underneath the wrapping paper and inside the gift box.

Your finances are irrelevant to acquiring a pet.

Although you do not need to be a millionaire to enjoy pet ownership, responsible owners undertake the care of their pets in sickness and in health. Young animals initially require frequent veterinary visits for the first series of inoculations, or vaccinations, against common contagious disease. These visits are invaluable because they protect the pet from illness and also give the owner the opportunity to discuss ways of preventing potential problems relating to health and behavior. Pets can get sick at any time and require veterinary attention. Unfortunately, health insurance for pets is not widespread and hospital costs are often the only obstacles to treatment. Even healthy pets require basic investment for upkeep. Pet food costs may not be a significant portion of your budget if you own a Maltese or a cat, but a big dog has a big

Chelsea Belle and Ruby are pretty, but more importantly, they are healthy and well cared for. (photo courtesy of Daniel Wallace)

stomach to fill. Grooming a German Shepherd Dog can be done periodically at home, but if you own a Persian cat that resents being combed or a Shih Tzu that looks like a mop within six weeks, you need to consider the cost of professional grooming.

If you are about to introduce a new pet into your home, it might be helpful to put some funds in a separate account for pet care. Most things that are worthwhile having are worth planning for. Acquiring a dog or cat should not force you to overextend your budget, your time and energy or your peace of mind. If you have any doubts regarding the emotional or financial commitment of pet care, you will never regret waiting a little while longer.

No special skills or experience are required to raise a dog or cat successfully.

Pet ownership does not have any educational prerequisites. Prior experience in anything is an advantage in life, but you have to start somewhere. Be realistic about your knowledge, and do not be afraid to ask for advice from people in the pet care industry. Make an appointment for an office visit with a local veterinarian. This can be a wonderful opportunity to dispel many popular misconceptions about pets and will establish a good relationship with your pet's doctor before doctor and patient ever meet!

Books can be helpful, but also confusing, depending on the subject you are researching. It is one thing to read about something in theory, but quite another to put it into practice. Begin Obedience training your dog early and research local training classes in advance. You need time, energy, patience, basic finances for initial and ongoing care, housing and the willingness to love. The desire to nurture and to protect children or their surrogates is one of the finer aspects of human nature. This instinct is not learned in a classroom. On the other hand, the quality of the care that we provide will be shaped by experience, through trial and error. There is much to learn about pet care and in the process, you will discover much about yourself.

The "pick of the litter" is the biggest one.

There are many criteria that define the "pick of the litter." The size of a kitten or pup relative to littermates has little or nothing to do with making a superior pet in any way.

First of all, your "pick" may not be someone else's "pick." We all choose our pets partly because of a history of experiences and personal impressions of what is a beautiful dog or cat, what temperament is ideal, and what

When selecting a pup, it is helpful to meet the mother and observe inter-action with littermates. (photo courtesy of Maureen Sneyd)

characteristics best suit our own private reasons for adopting a pet. The biggest puppy may be destined to become the most dominant in temperament and may not be a suitable family pet. The biggest Boxer puppy may have an even temperament and best represent the physical ideal of its breed, but if it is "cryptorchid" (a hidden or undescended testicle), it is undesirable for show or breeding. An unusually large Chihuahua might defeat the owner's purpose in selecting a Toy breed to begin with.

You may have a preference for a male or female. A male dog, for example, may be more dominant or aggressive than a female and may be rougher in play, depending on the breed. The breed's adult size, or the anticipated size of a mixed breed, will help to determine how much exercise might be required or how well the dog or cat will fit your family's needs. The breed will determine how much food, exercise and grooming is required over a lifetime. Your perception of what is physically attractive, such as coat color or size, is among the least significant considerations.

For most pet owners, the "pick of the litter" should be from the normal range of puppies or kittens. The most important criterion is temperament, but this is often difficult to judge in young animals. Thus, you probably should avoid extremes in any detail, such as an extremely small or large individual, or even an extremely friendly or cautious one. The best pet for the average pet owner is the average pup or kitten, both in size and in temperament. This rule of thumb could help to avoid some of the less than average problems later on.

The next consideration must be the animal's health. After you have selected your pick, schedule a veterinary evaluation. Health and a socially stable temperament must always take precedence over other super-ficialities.

The "runt" will be sickly and die.

The smallest or weakest pup or kitten in a litter is often referred to as the "runt" of the litter. The runt may be physically ill and, consequently, may be rejected by the mother. This behavior may seem cruel but it directs the mother's attention and energy toward healthier offspring that have a better chance of survival. The runt is not necessarily sick, but might simply be smaller than the other newborns. The runt is at a definite disadvantage if competition between littermates is keen. This might occur in large litters or if the mother's milk production is limited. If the runt's nutrition is supplemented or if a medical problem is identified and can be treated, physical development may equal that of siblings in time.

There is some evidence that suggests that the emotional and social qualities of a runt are the animal's greatest long-term disadvantage. The runt may be more anxious and fearful in general, and may be less socially interactive. In general, aggression may be more quickly and more intensely aroused. Still, many animals that were not the runts are easily frightened and aggressive in many situations. Many that were said to be the "runt" have grown into stable and trusted family pets. Considerations which may be just as important in a pup's temperament include the mother's temperament, her maternal experience and quality of care, the identification of nutritional deficits or health problems in newborn animals, and the skill of the human selecting the pet.

In the wrong hands, pets with perfectly desirable temperaments can become undesirable companions. Subtle behavior problems may be ignored or allowed to persist in the average pet. These may then evolve into long-standing and more difficult issues. Pets with more obviously challenging temperaments can reform and exceed everyone's expectations, perhaps because they demand and receive immediate training correction by the astute and motivated owner. For the average pet owner, the average puppy or kitten may be a better choice. The runt may or may not develop distinctive difficulties and could become a special pet for a special owner.

Hunting dogs salivate more than other dogs, and this is part of having a "soft mouth."

The salivary secretions of hunting dogs do not differ from those of other dog breeds. The saliva, secreted by specialized glands, helps to digest food, lubricate swallowing and keep the mouth clean. It also aids in self-grooming and hygiene. Grooming the young is of survival value and, directed toward another animal, it has an important social function.

Salivation is triggered by the anticipation of food (by sight or smell) and by the sensation of food (taste and texture) in the mouth. Salivation can also be induced by nausea or abdominal discomfort due to many causes. Some medications can increase or decrease salivary flow.

Dogs that belong to hunting breeds or those that are trained to hunt salivate for the same reasons that other animals do. Dogs that retrieve their owners' prey ideally do not mark or mutilate the prey when carrying in their mouths. This inhibited bite, an abbreviated predatory instinct, may trigger the dog's digestive processes. If hunting dogs salivate during cooperative hunting efforts, it is a natural phenomenon. It might also be due to mechanical overflow of saliva which is not easily swallowed when the dog's jaws remain open to carry a retrieved target.

Pet behavior problems are usually due to loneliness, so adding a second pet is the solution.

The introduction of a second pet in an attempt to resolve your first pet's problem behavior is like having a baby in order to save a failing marriage. Almost without exception, it is better to deal with any current problems before compounding them with additional ones. Even if the second pet has no inherent problems, your resident pet will be impacted.

Both cats and dogs require a period of adjustment to any change in the household. Preexisting behavior problems may be aggravated and new ones may appear during this transition. Resident and/or newly introduced cats and dogs may urinate or defecate inappropriately in your home, for example, and aggressive conflicts can flare. If you are already at your stress limit with one problem pet, why complicate your life further with the addition of another one?

In most cases where pet owners attribute a pet's problem to loneliness, the pet is perfectly content living without competition for food, water, toys or the owner's attention. As human beings, we are social animals and tend to project our unique emotional needs onto our pets. We might object to being alone, but we are not dogs or cats. Dogs interact with people much as they would with other dogs, and so in a real sense we fulfill their social needs (although every dog deserves a good time in the park with pals). Cats and dogs that have been raised alone often have difficulty sharing their home with another housemate. Loneliness is a sense of isolation that comes from a social void. It is a human emotion. Certainly, pets can and do suffer from social

Sara, fifteen years old, took a while to accept Hershey, two years old, but now they get along just fine.

isolation. In our pets, however, behavior problems commonly arise from many other sources. One of the prominent ones is a need for stimulation in the form of exercise, play and positive interaction with their owners.

The addition of a new pet might provide the resident pet with intellectual and physical challenges if there could be some guarantee that the two would bond instantaneously and forever. Some pets blossom with a companion but, should the match not be "made in heaven," you may be forced to make decisions. The well-being of both pets is jeopardized, and the repercussions will linger even if one of them is removed. Unfortunately, there are no guaranties in life. You are wiser to deal with what is known than what is unknown.

When adding another pet, a male and female always get along.

The only relatively predictable occasion for an immediately friendly encounter between a male and female is if the male is not castrated and the female is in full "heat." Even this union, however, may begin with some aggressive posturing. In cats, for example, mating lasts only seconds, and tension between the pair returns to drive them apart.

Bonds or alliances among animals, referred to as friendship in people, do exist, but the gender of the individuals is among the least important factors.

The single most recognized influence in determining whether two cats or dogs will get along is their respective experience with other cats or dogs during the formative phase of social development. Those that grow up in the presence of others of their kind often adjust better to new social situations later in life, but this does not mean that they always will. There are limitations to the ability to accept the presence of another housemate. Some animals raised with others never tolerate housemates and prefer to remain solitary. Those without prior experience can sometimes learn to tolerate the presence of another pet, although they may not become close companions.

This brings us to the second most important ingredient in the predicted interaction between animals, luck. Some cats and/or dogs that are raised together from a young age may never coexist peacefully, regardless of their gender. Some that coexist for extended periods of time undergo abrupt changes in their social dynamics and require professional intervention. The addition of a third cat or dog can disrupt the relationship between residents and lead to the formation of new alliances. A conflict between housemates can spontaneously erupt in reaction to something or someone threatening outside. A prolonged period of social upheaval may ensue.

Males can coexist and even develop close alliances. Females can interrelate in harmony. A male and a female cat and/or dog may live happily ever after, or they may grow apart. An older female cat or dog may have a more difficult initial adjustment to a male kitten or puppy, or they could have the equivalence of a winter-spring romance. Some pets seem completely mismatched and yet become inseparable.

When and if we uncover what determines a good "mix" we might have greater insight into our own interpersonal relationships. Conversely, once we more clearly define our own behavior we could better identify social predictors among pets. In the meantime, several suggestions can be made.

- Try to introduce an additional pet when both current resident and new housemate are still young (kittens or puppies).

- Make the introduction a very long and intentionally slow process.

- Restrict the newcomer's activity and delay any face-to-face encounter for at least one to two weeks while the current resident adjusts to the new one's scent and presence.

- Their initial introduction must be brief and strictly supervised.

- You might begin by feeding them at opposite ends of a room and then separating them after mealtime. Move food dishes closer together by several inches over many days.

- Allow the newcomer out to explore its new home for brief periods. Confine the resident pet so the newcomer's initial territorial patrols can be more comfortable.

Some tension is normal and some conflicts should be expected when cats or dogs are first introduced. In general, it is best not to interfere. In most cases, animals will evolve toward a stable social order of their own making. If fights are extreme at any time, the input and guidance of a veterinary behaviorist may prove invaluable. Although we use different terms to describe similar emotions and behaviors in other animals, it is understood that, in people and animals you can't hurry love nor can you force it to happen.

If aggression occurs between a newly introduced dog and the resident dog, they will never get along well.

Dogs are instinctively social animals and seek to establish alliances with other dogs or the next best thing, people. Some dogs are particularly antisocial, however, toward other dogs and people, too. Aggression during social intro - duction can be a sign of fear but it can also be a declaration of sorts, in which the individuals establish rank. A new dog introduced to a family will find a social position in the group's hierarchy.

The first challenge is likely to come from the most obvious source, the resident dog. Aggression can occur at their first encounter but in many in- stances, the two may investigate each other without apparent rivalry. As the resident realizes that this newcomer will be staying around for a while and as the new dog begins to adjust to the rhythm of the new home, social issues of relative rank need to be resolved. After an initial period of truce, subtle or pronounced aggression can develop between the dogs even after consider- able delay.

It can often be helpful to introduce the dogs on neutral ground. Dogs that are used to living in their own home without canine competition may resent the intrusion of a stranger in their living room or yard. The new dog may be perceived as a territorial threat and not as a potential playmate or ally.

If they first meet in a local park or playground to which neither has any claim, they can start out on the right foot, or paw in this case. Once the new dog is brought home, feed each dog in separate areas so that competition over food does not become a source of conflict.

Make sure that each dog is given adequate attention and does not need to resort to undesirable behaviors to get it. The resident dog will be particularly uneasy during the transition period and deserves extra attention. An extra walk alone with you or an extra game of ball in the park without the new dog are helpful. Make a point of spending time petting or brushing your first dog in the presence of the newcomer to reinforce that competition for your attention is unnecessary. This will also teach the association of a positive outcome with the new dog's presence instead of a negative one.

Although you may try to give your first dog priority, canine society is not a democracy. The younger and stronger dog may successfully challenge the other for higher social status. In most cases, this is accomplished in subtle ways and obvious aggression may not be noticed. Males will frequently dominate females although they may appear to defer to each other to the untrained eye. Most dogs will coexist well, given time, although there are always exceptions. A particularly dominant male dog, for instance, may never accept the addition of a second dog, especially a male. Rivalry between males is among the most intense and difficult to control. If you recognize this behavior in your male dog, a female would be a wiser choice as a second pet. Aggression between females vying for social prominence does occur but rival males are often, but not always, more dangerous.

If aggression occurs between a newly introduced cat and the resident cat, they will never get along well.

Aggression between two cats, by one or both, is common at first meeting. Their initial reaction, however, is not a reliable indicator of their relationship once the dynamics have had time to stabilize. The two cats may eventually become close companions, or they may remain steadfast in their hostility.

Most of the time, new housemates will at least learn to tolerate a shared living space. In most cases, some form of aggression is seen and is considered normal, even in cats that are used to living with other cats. Cats raised with other cats have a better chance of learning to live with one another. Those that were raised as single pets may need more time to adjust, and the transition may be more difficult.

Exceptions can be the rule, however, and some cats seem to gravitate toward each other regardless of their past histories. This is true whether the individuals are of the same or opposite sex. If the individuals are gradually introduced, chances are better that they will eventually live well together. If their first meeting is rushed or forced, they may never forget the trauma of their first encounter and forever associate each other with the initial emotional turmoil.

Unless the aggression is extreme, the cats should be allowed to resolve their conflict without intervention. If their interaction does not improve, and is severe or of any concern to the owner, it might be necessary to consult a veterinarian or veterinary behaviorist for additional help. Occasionally, individuals never come to tolerate each other, and conflicts continue or even escalate. Like people, individual cats and dogs do not always develop an affinity for each other and require indefinite separation. If newly introduced animals have not begun to accept each other after roughly three months and if life is still uncomfortable by six months, it might be time to seek professional advice or to place one of them in a new home.

The best way to introduce two pets is to confine them together in a room and check on them later when things are quiet.

This extreme practice is almost never necessary nor is it recommended when introducing pets for the first time. Slow, steady and supervised meetings are almost always better than coerced confrontation. Animals that might otherwise have adjusted well to each other may react so intensely and adversely that their mutual association with the damaging memory is difficult to undo.

It is frequently helpful to isolate the new pet from the other in a quiet and comfortable room. The newcomer can adjust to your private visits to play or deliver meals. Your new pet can learn to feel secure in a limited space before continuing to explore the rest of your home and before being burdened with the additional stress of learning to interact with the resident pet.

This method works best with cats or when introducing a young pup to a resident dog. It is more difficult to isolate an adult dog that is being introduced into your household, but the idea is to give each dog as much space as possible and to provide them with a secure basis for a relationship with you without the presence of a competing house mate.

One of the ways to plan their first encounter is bring them together at mealtimes and only for brief periods. Return your new pet to the room afterward so that it can regain a sense of security. Feed them at separate ends of the same room when you are present to supervise. Eventually, you should be able to allow them more freedom to approach each other. Some aggression may occur, but in most cases, tension will soon subside. In rare situations, forced and inescapable encounters become a last-resort solution if other methods of incorporating new pets or re-establishing disrupted dynamics between housemates have failed. Even so, this technique is risky at best and should be conducted with strict guidelines provided by a trained veterinary behavior specialist.

If you do not own a house with a fenced yard, you should not have a dog.

Whether you own a home or have a backyard is not a prerequisite to pet ownership. The most important thing that a dog needs to ensure a quality life is a committed owner.

Social interaction in the form of play and petting, exercise during at least two walks every day, good food and clean water, and emotional and physical security are the basics deserved by every pet. If you happen to own a house with a yard, it is a convenient, additional space for a dog. In any event, your yard should not replace taking your dog for daily walks. In some ways, relying on a yard for your dog's exercise and recreation is similar to relying on your television to baby-sit your children. It might be all right as a temporary entertainment device, but nothing beats the real thing—you. Having a yard might even be a handicap in your relationship with your dog if it means that you take your dog for granted and become slack in the attention and direction you give. Regardless of the size of dog or the size of your home, dogs need more exercise and stimulation than any yard can provide.

Large breed dogs should not live in apartments.

If your apartment is big enough for you, it is probably big enough for your dog. If you think about it, the size of your home frequently reflects your lifestyle. Active people spend little time at home and have many interests and activities outside the home. The same should be true about your dog. If you make certain that your dog has enough exercise, like daily walks and playtime, that is appropriate to age and temperament, the floor space of your home is relatively unimportant. Both you and your dog, whether a Schipperke or Rottweiler, can play, exercise and socialize with other dogs or people outdoors. By the time you return home, you will both be contented to rest quietly or take a nap!

When you think about it, many of us only come home to sleep at the end of a long working day. We desire larger homes as we acquire more possessions, expand the family, earn a higher salary or simply want to make a social statement. The size of your home has little to do with how well the space can accommodate a large pet. A Great Dane can be content in a one bedroom apartment if it is well cared for. On the other hand, a tiny Yorkshire Terrier could live in a ten-bedroom mansion and never have basic needs for social interaction and daily exercise met.

A minimum of two leash-walks and additional quality time with you, should leave the dog content to rest when left at home. A sleeping dog does not need much room. Of course, some dogs are just clumsy in tight spaces. A happy Dalmatian can have a whip for a tail and knock over ornaments on your coffee table. A bouncing Boxer may have a docked tail but could be so determined to greet you at the door that any obstacle will be dumped over or shattered. This might be frustrating but just remember—your pet is happiest wherever you are.

Smaller dogs are less aggressive and make better pets for families with young children.

The adult size of a dog is an important criterion in the choice of a family pet. In general, small dogs eat less and require less space to exercise compared to larger breeds.

However, a dog's need for time, attention, basic veterinary care, grooming and Obedience training is identical regardless of size. When selecting a pet, parents must consider their child's emotional maturity and physical abilities. If a child tends to be a bit rough, a medium- to large-sized pet might be sturdier. Walking a dog of almost any size can be challenging for children if the dog pulls on the leash. For more delicate children, small and Toy breeds may certainly fit the bill. There are many delightful small breeds. The Pug and Shetland Sheepdog are consistent and deserving favorites.

Do not be fooled by the cuddly, cute and compact dog. Small dogs frequently become more aggressive than larger ones, perhaps because they are more indulged during formative and impressionable stages. Owners of small dogs tend to omit Obedience training which is invaluable in shaping any desirable behavior.

Small dogs tend to be more sheltered in general and are frequently walked less than larger dogs, particularly in less comfortable weather. They may be exposed to fewer people, dogs and a variety of experiences. Consequently, they can become more fearful, more anxious and more easily startled into defensive aggression. If you make the decision to raise your small dog as you would a large dog, the size of the pet will be outweighed only by the pleasure it will bring to all your family.

Probably the greatest difference between small and large breed dogs is, quite literally, their size. Aside from a very few breed-related characteristics (Terrier breeds in general are more dog-aggressive as compared to Siberian Huskies, for example), it is unlikely that dogs are even aware of their size. Think about the Lhasa Apso that terrorized you for years and then remember the Doberman that was afraid of clouds in the sky. What about that well-behaved Irish Setter that practiced Obedience skills with the owner in the park? Remember the Toy Poodle who discovered that biting was an effective way to say "please"?

Small breeds like Minia-ture Pinschers should be Obedience trained, just like big dogs, to become good family pets.

A large dog can be born with a gentle and placid disposition and make a wonderful family pet. Many dogs are terrified of the energy generated by healthy kids but small breed dogs may be additionally prone to unwanted attention because they are "kid-sized."

Of course, dogs probably have a limited concept of their own size, which does influence them. Young dogs get social feedback about themselves from interacting with littermates, other dogs and people. A small dog with a domi-nant temperament will be reinforced for unacceptable behavior if the owner does not appreciate the art of canine social behavior (see chapter 7, "Aggression"). Dominantly aggressive dogs of any size soon recognize that kids are easily dominated and can be less tolerant of perceived insubor-dination in children.

The average medium- or large-sized dog receives more Obedience training from the average owner as compared to small- or miniature-sized dogs. This means that two dogs of equal temperament (but contrasting size) have the potential to become a socially acceptable pet or an undisciplined one raised without standards of desirable behavior. The size of the dog is the physical package, but the spirit of the dog is what determines suitability for your family. *You cannot change the outside, but you can influence behavior with appropriate and early training.*

Cats are women's pets.

Real men do wear pink, eat quiche and own cats. There is nothing effeminate about having a pet cat. Cats are alert, inquisitive, responsive, athletic, affectionate and graceful. It is not that cats are independent of people, rather that dogs relate to people more conspicuously and rely more intimately on them to satisfy certain social needs.

If a prospective pet owner, man or woman, wants a pet for hiking company, then a cat may not fit the bill. If someone's definition of an ideal pet is a friendly face to come home to that does not require long walks in the rain, then a pet cat would be perfect. Pets mirror our social facades but also suggest our more introverted natures. Truck drivers can adore Chihuahuas, and women can handle Chesapeake Bay Retrievers. A petite, shy and self-conscious woman can feel more confident, secure and sociable through her Labrador Retriever or Beagle. A football-playing, weight-lifting man can express his more pensive side by curling up with a cat and a good book. Our choice in pets reflects not only who we are but who we aspire to become.

Cats make good pets for busy people because they require no attention and little care.

Cats do make good pets for people who lead active lives but because they thrive without human interaction. Domestic cats are instinctively self-sufficient creatures. They evolved as solitary hunters with only minimal social contact with their own kind. Some say that much of a cat's social interaction with people mirrors that with its mother, or at least that is one interpretation. Like the queen who cares for her kittens, we supply food, grooming and hygiene. Even petting has been compared to the action of a mother cat's tongue on her kitten's coat. Compare this with the ancestors of the domestic dog that were pack hunters and lived in close cooperative units. Today, dogs relate to their owners much as they would relate to canine pack members.

The process of domestication required the selection of individuals that were most likely to tolerate people in closer proximity and in roles that would have been intolerable to the wild individual. The domestic cat, however, has not received the same careful attention to specific physical or behavioral traits that produced the multitude of dog breeds. Many cats would revert back to fending for themselves if necessary. Pet cats have been altered little by living with people. This might explain, in part, why many cats do not depend more on human care. We can only cautiously compare dogs to cats, when discussing their relationships with people, because their natures are different.

The domestic cat is not derived from a socially interactive and dependent animal as is the dog. Cats are a more subtle creature in many ways. Much of their friendly behavior is understated, but their sincerity is not less than a dog's. A cat's greatest sign of affection may be to shadow us as we walk through

Cats are affectionate pets that can amuse themselves, but most would rather play with you.

the house, keeping close by when we sit, and brushing their bodies against our legs. A dog's delight is more likely to be conspicuous, with bounding leaps, excited whining, slobbering kisses.

Cats do not need to go out for walks (unless their owners are inclined to try it for additional recreation) and, because they are sprinters, need little exercise to keep physically fit. Confined as house pets, cats rely on their owners for much of their daily physical and intellectual stimulation. Indoor and outdoor cats need to interact with their owners although individuals vary in their desire for close human contact. Some cats are less inclined to play and prefer brushing, or vice versa, yet they still require the daily commitment of a devoted owner.

We are all overextended and live at a hectic pace. Our pets help to steady our psychological balance when we find ourselves socially isolated from our own kind, or perhaps, because of our own kind. Whether we share our home with cats or dogs, they remind us that real success is reflected in the happiness of loving eyes.

Siamese are more vocal and intelligent than other cats.

Few will argue that the Siamese cat has a distinctive voice. Some have compared it to the cry of a child, so clear and pronounced are the pitch and tone. This does not mean, however, that Siamese vocalize more than other cat breeds. It only proves that when a Siamese cat calls, people will listen. The Siamese voice is an instrument that demands attention but no scientific study has ever been undertaken to show that they vocalize more frequently than other cats.

The vocalization of cats deserves more study but we do know that the amount of vocalizing and the type of calls from an individual cat, regardless of breed, are strongly influenced by experience. If a cat's needs (e.g., for attention, food, exercise) are met, there may be no motivation to learn ways of attracting the owner's attention. If a cat one day vocalizes because of a lack of stimulation, and the owner offers food, that cat will learn a valuable association between the vocal behavior and the owner's response. Siamese are no more intelligent than other cats. (Please refer to the discussion regarding intelligence and dog breeds that follows.)

Intelligence in dogs varies according to breed.

The definition and evaluation of human intelligence is the subject of ongoing debate and controversy. Hundreds of psychometric tests are available to quantify and qualify the abilities of young children, adolescents, adults and senior citizens. To date, there is no intelligence test that can be adequately or convincingly applied to either dogs or cats. Indeed, the discussion continues not only with regard to the intelligence of nonhuman animals but as to whether they are capable of thinking at all!

We are not the only intelligent life on this planet. Other animal species have functional brains not so different from ours and likely have the experience of "thinking" in their own ways. Their thought processes must be a function of their central nervous systems and the demands necessary for survival in their natural habitat. Any human evaluation of intelligence in nonhuman animals will be necessarily influenced by our perception of desirable qualities that constitute intelligence.

The creation of dog breeds over thousands of years of selective breeding has resulted in dogs that vary widely in body shape and size. The intended functions of individual breeds, as well as some unanticipated but useful abilities, were then emphasized through the process of domestication. Golden Retrievers can retrieve, but if Goldens cannot herd sheep are they stupid? Bloodhounds can be invaluable in search and rescue operations, but if they cannot pull a sled are they intellectually challenged?

We are far from being able to reliably and methodically assess the intelligence of dogs according to breed. "Insight" learning reflects an individual's ability to interpret or cope with an unfamiliar situation. A simple test to evaluate an individual's problem-solving skills may be of greater value: If you are walking your dog on a leash, for example, and your dog passes on one side of a telephone pole and you on the other, does your dog keep trying to pull forward or back up, turn around and walk on your side of the pole? If your pet is trying to reach a toy that has rolled or been pushed under a piece of furniture, does the dog make attempts to reach the object from different angles or another side of the furniture? To date, no scientific conclusions can be made regarding dog intelligence according to breed or any other criteria.

It is likely that we will some day find that basic intelligence, in dogs as well as other species, has little to do with breed or race, but with an individual's innate abilities and the environment.

Border Collies are the smartest breed.

The issues surrounding canine intelligence are complex. The association of any given dog breed with a higher or lower measurement of intelligence is unfounded. The Border Collie is a lively dog in mind and in spirit, but to proclaim its superior intelligence is akin to declaring chocolate to be the best flavor of ice cream (or frozen yogurt, if you prefer).

The Border Collie, and other Herding breeds, are without question well suited for protecting their flock and cooperating with their owners' signals. These dogs are tireless, rugged, agile, determined and willing to please, but so are many other dogs. Many breeds that work in some way (hunting, herding, etc.) would fit this description. These qualities can be assigned to almost any breed of dog by those who admire them. A dog's innate or learned talents will be highlighted if the animal is placed in the right situation or given the opportunity. In the hands of some people, the Border Collie may not have a lifestyle suitable to the breed's physical or emotional needs and could develop into a neurotic and undesirable pet.

In a less than ideal environment or circumstance, any dog (or cat or person) might appear less intellectually gifted. Intelligence, much like beauty, is influenced by the eyes of the beholder. What you consider to be an intelligent behavior may not coincide with my interpretation. Your notion of what makes a pet desirable may not match my needs as a pet owner. Who was more intelligent, the tortoise or the hare? Is the swift-footed Border Collie brighter than the stocky Bulldog? One thing is certain. Your pet, like your children, will always be the most beautiful and talented. No one should try to convince you otherwise.

Afghan Hounds are the least intelligent breed.

The Afghan Hound is one of the most ancient of dog breeds. It was bred to be a desert sprinter, complementing the hunting of many admirers in the Middle East. Few Afghan Hounds these days practice the purpose for which they were intended. Once in a while, their tremendous speed and graceful athleticism are showcased in circus acts. The Afghan requires daily grooming to keep a long and flowing coat untangled and attractive. Perhaps this quality contributed to their image as the dumbbells of the dog world. Many intelligent and hard-working men and women give priority to their appearance and enjoy being well groomed.

The Afghan is a sighthound of high spirit and keen senses. Any unkind reputation may be due, in part, to its former association with society's elite and the resentment of the less socially powerful. The Afghan may have been owned by people who were perceived to be hedonistic, pampered and vain, but that should not detract from an appreciation for the natural beauty of the dog and its regal place in the canine kingdom.

Chocolate Labrador Retrievers are easier to train than Labs of other colors.

A dog's coat color has never been linked to trainability or the lack of it. Chocolate Labs are beautiful but so are the black and yellow ones. It is important to keep in mind that physical attributes are less important than temperamental qualities. We should all rely less on our personal biases regarding the outward appearance of pets (and people) and look beyond to the being within. The selection of a dog according to coat color is an acceptable criterion as long as it is not allowed to "color" the dog's temperament, health and suitability for the prospective owner's lifestyle and experience.

Sheepdogs relate to their owners as they would to their flock.

The instinct to herd can be strong in shepherding breeds such as the Shetland Sheepdog, Maremma and Border Collie, although individuals of many other breeds can show this inclination, too. The Rottweiler, for example, was originally bred as a cattle dog.

The herding instinct can appear even in house pets, but will be directed toward less appropriate targets. For example, the herding instinct can be directed against moving vehicles, children in the park, people on bicycles or other dogs on the playground. The dog does not relate the same way to owners as to sheep or cattle although behaviors related to the herding sequence (such as stalking, nipping at heels, chasing) can emerge in play and during unrestricted playful exercise.

The herding instinct can be developed in many ways. One is to raise the pup with the flock so that the dog actually relates to sheep as members of its pack. The dog can then be expected to naturally defend the pack against intruders or other predators. Some pups have such gifted herding ability that they instantly round up a "herd" without prior contact or experience. This is one way that shepherds or ranchers evaluate which pup in a litter is best suited to the task. The herding pup's inborn talent will require little besides training to respond to specific whistles or calls that direct the dog's work with great precision even when the owner is far away.

Some dog and cat breeds are "hypoallergenic."

There is no such thing as a hypoallergenic pet. A person can be allergic to certain breeds and suffer no symptoms when exposed to other breeds. Some people can even be allergic to individuals of a particular breed yet have no discomfort with other individuals of the same breed. Sometimes allergies develop only after a period of exposure to a pet. Even if a pet does not trigger an immediate allergic reaction, an owner can develop an allergy weeks or years later. It is also possible that allergic reactions to a pet subside with time. Dog breeds such as the Poodle and Bichon Frisé, as well as some members of the Terrier group, such as the Airedale, are less frequently associated with allergies in people, but there is no guarantee that an immediate or delayed sensitivity will not appear.

The most common symptoms of pet allergies in people are related to the respiratory system. Upper respiratory signs such as itchy or burning eyes and nose, sneezing and coughing can sometimes progress to more severe reactions. Lower respiratory signs can affect the breathing passages leading to the lungs, resulting in bronchitis or asthma. Allergies can start subtly and evolve to more serious symptoms, or flare suddenly and intensely. They can become chronic and nagging, such as ongoing sinusitis. Signs of allergy can also take the form of skin problems. Direct contact with an allergy-triggering pet may cause mild symptoms such as local itchiness on the skin's surface. Allergic skin reaction can range from small localized bumps called "hives" or to generalized eruptions over the body.

If you are allergic, your only option is to place your pet in another home.

If you have a severe allergy to an individual pet or breed, it is important to consider your health first. Allergic reactions can seriously impact your health and can even be fatal. Even if your allergic symptoms are mild, your discomfort can interfere with the enjoyment of your pet. Your dog or cat will not understand why you seem withdrawn and avoid contact. Consider what is best for you and your pet.

If you have allergies to dogs or cats, speak with your physician about treatments that can alleviate or control your symptoms. Desensitizing injections can eliminate signs of allergy permanently or temporarily, but they may also fail. Visit a breeder's kennel and spend time with an expectant mother even before the litter is born. This will allow you to test your tolerance to a breed without falling in love with a kitten or puppy that must later be placed elsewhere. When you select your pet, spend at least half an hour or more to be sure there is no obvious problem.

It can be helpful to restrict a pet's access to certain areas, for example, the bedroom. Wash your hands after you have touched your pet. Keep your skin moisturized. Keep your home well vacuumed and ventilated. Keep your pet's coat well groomed and clean. This does not mean that you should bathe your pet daily or weekly. Most dogs need not be bathed more than three or four times a year. An indoor cat rarely requires a bath although some allergists advise weekly bathing to help remove the allergy trigger on a cat's coat. The problem with this recommendation is that it may create dermatitis (skin infection) and predispose the cat to other infections. It could also cause the cat to groom itself even more which might defeat the purpose of bathing. Allergies to cats are due to a sensitivity to a protein in their saliva, which is deposited over their body surface as they groom themselves. Bathing may dilute the allergy-causing substance on the cat but does nothing to control the substance that is shed in your home and which remains stable for extended periods of time.

Dogs have many different coat types. Some breeds shed less than others or have an oilier coat, but every dog sheds some. In fact the shorthaired breeds may shed more than some longhaired breeds. Individuals within a breed can also have slightly different natural oils or odors or coat quality. It can be helpful to use a dry shampoo to absorb excess oils or odors from your pet's coat. Passing a dryer sheet over your pet's coat can also be helpful. If you have had allergies to pets in the past and would like to adopt a cat or dog, it is probably best to avoid the species or breed that gave you problems.

There is less variety in hair quality between cat breeds, although any difference can be significant to a person suffering from allergies. Cats such as the Persian have fine, very long coats that shed in addition to requiring daily grooming. Breeds with short, sparse coats such as the Siamese or Burmese may cause less problem. Cats with even finer coats, such as the Rex, shed very little, but some allergy sufferers are still sensitive to this breed.

The age at which a pet is acquired is not important.

All animals are a product of individual genetic traits and their lifetime of acquired experiences. For dogs and cats, predisposition toward certain

The choice of a pet often reflects and reveals who we are.

temperament qualities such as sociability and tolerance to handling is open to influence to some degree.

The greatest period of flexibility with regard to social behavior occurs in puppies approximately six to fourteen weeks. In kittens, this critical period of development is not as clearly defined. It likely coincides with the age range observed in pups although it may begin and end at a younger age. Young pets that have had unpleasant encounters or that have been socially isolated from people during this impressionable phase will have a more difficult time to unlearn their experience (or lack of it).

On the other hand, many pets have been acquired as mature adults and quickly mesh with the fabric of their new families. The match between pet and owner can be so comfortable, it is as if they had always been together. There is much to be said for the comfort and ability of devoted and patient owners. If an adult pet has had even limited exposure upon which to build more normal relationships with people, all that might be required is the opportunity to practice. Not every dog or cat that has had an unhappy or late start in life can become a good pet, but many can.

Stray cats often adapt to confinement as house pets even without prior experience, leading many owners to believe that they must have belonged to someone at one time.

Some pet owners feel they would like to adopt a puppy or kitten because they do not want to deal any sooner than necessary with the death of a pet, which might occur when acquiring a slightly older dog or cat. Yet there are no guarantees. Young animals do not always survive to old age and pets adopted as adults can live long lives with proper care.

The advantage of adopting a young pet is that you are beginning essentially with a blank slate, although its basic nature is genetically programmed. A younger animal will have the opportunity to learn desirable behavior in the context of your home. Any problem behaviors that might develop can be traced. When an older animal is adopted, previous socialization (or lack of it) may be unknown, and many behavior patterns are already formed. The origin of any problems may never be known. This does not mean that problems will be irreversible, only more of a challenge.

Animal shelters are filled to bursting with extraordinary creatures who deserve a chance to reach their potential.

Pet store puppies are no different from dogs bred by breeders.

Pet stores have received much attention for poor housing and medical management and some have tried to improve that. One of the problems that remain with pups purchased from pet stores is that they may remain in small cages until they are several months old or more before being purchased. This predisposes them to upper respiratory infections, for example, because of close contact with other animals and limited ventilation. This also prevents many pups from being properly socialized to people. Other behavior problems that include difficulties in toilet habits may crop up. Although many pet store workers do their best to interact with the pets, they cannot devote the same quality time that a breeder usually does.

Pet stores may acquire their "stock" from local backyard breeders as well as "puppy mills" that mass-produce dogs of many breeds. These are dog farms, mostly located in the Midwest that continue to supply pet stores with pups. Exposed by the media and local animal lovers, they received notoriety for abusive housing conditions and mistreatment of their breeding stock and pups, and the worst offenders were shut down by law enforcement. Although current operating conditions have not received attention recently, puppy mills continue to prosper owing to the demand for pet store puppies. Just because a pup has AKC or UKC registration papers as a purebred does not make it healthy or a quality pup for the breed.

The price of pups at pet stores is often inflated to cover store overhead. Falling in love with a pet store puppy has led to many sales based on impulse-buying despite the price tag. The importance of exercising caution and avoiding spontaneous decisions when purchasing a pet cannot be overstated.

Children and pets need to be taught mutual respect and they need to be carefully supervised to make sure they will not harm each other. (photo courtesy of Marva and Oscar Moxey-Mims)

5

Pets and Children

The best way to teach the facts of life is to let your children observe the birth of puppies or kittens.

The care of a pet dog or cat is a privilege, not a right. Control over the destiny of others more vulnerable and dependent than we are does not give us the authority to exploit their affection or functions.

Children can learn sex education in the classroom, from a book or video and through discussion with supportive and attentive parents. Dogs or cats are not science experiments for home use. Each is an individual with emotional and physical needs and relies on us for shelter and loving care. A female dog or cat does not need to experience pregnancy or maternity to improve the quality of her life. A male dog or cat need not prove his masculinity by mating. Rather than add to the tragic statistics of the pet over-population, have your female male pet neutered.

Treat your pet as a cherished companion and worthy canine or feline member of the family. The example that you set as a responsible and considerate pet owner and parent will teach your children what they really need to know about the facts of life.

Children will learn to be responsible by caring for a pet.

Regardless of how reliable and mature your child is, it is simply unrealistic to expect a pet to be a teaching tool. Parents should have realistic expectations of their own children. Children can be quick to lose interest in new toys, for example, no matter how much they pleaded with you to purchase it. Before you acquired a pet, they also promised they would cherish it forever. New friends, more homework, additional extracurricular activities will draw time and energy away from the family pet.

Your dog or cat's emotional and physical needs, however, are consistent and do not fluctuate according to your work or your child's schedule. Your child may not relate much differently toward a thinking and feeling dog or cat than toward a toy. Even if a child participates in feeding, grooming, exercising and playing with the family pet, a wise parent is aware that a child's interest and focus of attention will naturally waver and wander. Ultimately, owning a pet will not be your child's responsibility, but yours.

We each respond to a pet on different emotional levels and in unique ways. The symbolic role of a family pet can fluctuate between playmate, sibling and even surrogate parent for your child. Children should not be expected to parent anything or anyone prematurely. You can expect your child to behave childishly. You are the parent and can expect to supervise and uphold most of the pet's care. If your child fails to perform his or her expected share as care giver, you and your pet will bear the consequences. Before plunging your family into the lifelong commitment that every dog or cat deserves, see if your child is consistently successful at assigned daily tasks, like school homework or assisting you in household chores. Children should not learn to be responsible by caring for a pet. They should already be responsible before you acquire the pet by learning to respect and care for themselves.

Children are naturally kind to other animals.

Children and adolescents must contend with a confusing and aggressive world. Youngsters are capable of complex emotion and behavior that reflect where and by whom they were raised. Troubled kids may have few tools with which to communicate their own stress or grief except in the form of socially undesirable behavior.

Unfortunately, the family pet can be an easy target of turbulent emotions. Pets endure juvenile catharsis more often than is currently reported. Concerned parents should be vigilant of any abusive treatment of a pet. If your child seems to behave roughly with your pet or appears to resent the animal in any way, an underlying psychological issue may be the source. Psychological counseling may be worthwhile if you are unable to communicate with your child or prevent further abuse of your pet.

Children can inflict pain unintentionally on others, too. Youngsters can simply be unaware of their own strength or ability to physically harm a trusting pet. *Every child should be taught how to interact with animals.* A child must be guided toward desirable public conduct and needs the same instruction with pets.

Guidelines regarding when, where and how children spend time with the cat or dog should be established by parents and practiced on a daily basis with kids. For example, children must be taught to keep a distance when the dog is enjoying a favorite treat or not to disturb a sleeping cat. They should be shown how to gently caress the pet and warned not to pursue if the cat or dog tries to get away. Maturity is a recognition of our responsibility for the effect we exert on the world around us. That takes time to develop.

A good dog or cat has endless patience with children.

Very few parents have endless patience even with their own children. It is unfair to expect this from pets.

Children need education about how, when and where to interact with the family pet. Toddlers in particular should be guided, during supervised daily interactive sessions with an adult, on how to caress the pet dog or cat or both. A child must learn which areas may be petted and where the pet may

Even a good-natured dog will have limited patience.

resent being touched. The child should be guided on the firmness of his or her own touch.

There are limits to a dog's desire to cooperate in a game of cowboy. There is only so much "dress-up" that a cat can endure. Limits should be set so that the pet is left undisturbed during rest or feeding.

- Do not assume that your child is blameless in every situation.

- Teach the child to avoid touching when the pet is asleep in a favorite spot.

- Insist that your child not pursue if your dog or cat runs away.

- Teach your child that dogs or cats also have control over their bodies and that "no" means "no," except a pet expresses it a bit differently.

Every pet has individual limits of tolerance, patience and endurance. Ideally, these should never be tested.

Children should grow up with pets from a young age.

Growing up with pets is not a requirement for a happy childhood. Children may not develop an interest in a pet until their adolescent years or beyond. Many people who care deeply for animals do not acquire pets until later in life. Some children are afraid of animals, but that alone does not justify pet ownership. Adopting a pet that might otherwise be unwanted, except to teach the child to overcome its fears, could complicate the problem for everyone.

A fear of animals might respond better to your reassurance and the child's gradual exposure through illustrated books, informative documentaries and videos, and through visits to a pet store, the zoo and your neighbor's friendly dog and cat. All these entail far less unnecessary commitment until your family is truly prepared for pets.

Parents may be unable or unwilling to give priority to the investment of additional money, time and energy that any family pet needs. They should not feel guilty that they are somehow depriving their child in any way. Indeed, any reason to delay the decision to adopt a pet is wise.

The care of a pet is a lifetime commitment that must not be taken lightly. The choice should be seriously reconsidered if a child is not mature enough to appreciate a dog or cat; if a parent is unprepared to accept the restrictions on lifestyle that accompany responsible pet care; or, if any family member strongly objects to having a pet for any reason. Children can develop into nurturing and kind people who care about the planet and all its inhabitants without the experience of family pets—or with them if those pets are completely welcome.

A woman who is pregnant or considering having a child must avoid all contact with cats.

Toxoplasma is a microscopic parasite consisting of a single cell. Toxoplasma can infest most species of mammals, including sheep, pigs, horses, dogs and cats, as well as birds. People can ingest the parasite in infested meat or by accidentally swallowing Toxoplasma eggs. If a pregnant woman contacts Toxoplasma eggs and touches her face or mouth before washing her hands, she could contract the parasite, too.

A minority of cats carry the parasite but most have no symptoms of disease. Silent carriers can shed parasitic eggs in their stools and the litter box, for example, which can become an additional source of contamination for people. It is thought that cats become infested by eating contaminated meat or by frequenting a location where an infested animal has defecated. If a cat becomes clinically ill with the disease called Toxoplasmosis, symptoms can range from gastrointestinal to neurological. Although these can be severe, the patient usually recovers with treatment.

Toxoplasma is not a usual parasite in people, and most individuals that are exposed do not become ill with the disease. If ingestion occurs, however, the parasite could migrate into parts of the body that are vulnerable and less easily defended against infection in general. An unborn child is particularly vulnerable whenever the mother has any health concern. Massive exposure to Toxoplasma can cause miscarriage, blindness and mental retardation. *It is important to protect your pregnancy but also to understand that your pet cat poses only a minimal risk.* Less than one percent of pet cats actively shed eggs in their feces. Realistically, a pregnant woman has a greater chance of contracting Toxoplasma by gardening without gloves or from a child's sandbox, where roaming cats may have defecated.

Cats raised exclusively as house pets with regular veterinary care are unlikely to present any type of parasitic infestation. Outdoor cats are at higher risk, however, particularly if they hunt mice or other small prey infected by this parasite (and other illnesses). Regular fecal samples, even for indoor cats, are important because the parasite can lay dormant in carrier animals and infection can be reactivated by stress. Pet cats should not be allowed outside where they can hunt and consume infected birds or rodents.

As a precaution, pregnant women should avoid contact with the litter box. Ask your mate or friend to take care of the litter box at least until the baby is born. If no one is available to help you and if contact with the box is necessary, wear protective gloves. Wear gardening gloves during contact with soil or sand. Wash your hands after petting or brushing your cat and periodically throughout the day.

If you are pregnant or considering pregnancy, ask your obstetrician-gynecologist to draw your blood and test your level of antibody specifically against Toxoplasma. The presence of antibodies in your system does not mean that you have the disease. Instead, your antibody levels show that you already have a defensive system in place to protect against any accidental contact with the parasite. If your blood test indicates that you have no existing defense (zero antibodies) against this parasite, it is all the more important to protect yourself against possible infection. There is no point in testing a cat for antibody levels because the animal could test "negative" (zero antibodies) and still be an asymptomatic carrier. Basic hygiene and common sense will allow you to enjoy your pet and anticipate the birth of a healthy child.

It is safe for children to pet a dog who is walking with the owners.

A dog can actually be more aggressive toward strangers when in less familiar territory, even when with the owner. Perhaps this is because the instinct to defend other "pack" members is heightened by anxiety on unfamiliar ground.

We should all approach strange animals with caution. Regardless of how much your child is attracted to pets, and even if the dog is accompanied by the owner, teach your child to ask permission first. Some dogs seem friendly but do not interact well with children. Some dogs are uneasy with strangers. Some owners prefer that no one touch the dog for any number of reasons.

Prevent your child from running instantly to any animal, regardless of whether the owner is present or nearby. Teach your child to slowly, calmly and cautiously approach a dog it would like to touch. If the dog is distracted or if the owner seems rushed, leave them alone. If either the dog or the owner seem unfriendly in any way, keep away and direct your child's affection toward more welcoming and safe subjects.

If your pet has never been aggressive toward children, it will not endanger your infant in any way.

Although your pet seems fine with other people's children, the social impact of a new member in your family is vastly different for the family pet. Even if your dog seems harmless and happy to play with kids in the park, these youngsters are not steady competition for your attention or food, territory or social position.

Even after you have children, remember your first "baby" and keep your commitment to your pet. (photo courtesy of Estelle Schwartz)

If your dog predated the arrival of your child, the dog may view your child as a subordinate with lower social rank. Even if your child is older than your dog, canine social development is more advanced at an earlier age. Your child's leniency may actually signal submission and justify the dog's claim to dominant social rank. Dogs who have established "seniority" or have socially dominant temperaments may eventually perceive children as their potential rivals despite the fact that kids typically do not present authoritative figures.

Spontaneous displays of human affection common in children can be interpreted as challenges to the social position of the dominant dog. For instance, a child that affectionately throws its arms around the dog's neck could be viewed as an act of insubordination by your dog and could prompt punishment in the form of dominance-related aggressiveness.

Parents should be particularly careful to supervise their children when they reach the toddler stage because it is at this phase that dog attacks against children are most common. It is thought that dogs may have difficulty in recognizing the toddler and the cradled infant as one and the same. Attacks against toddlers are presumed to indicate the arousal of previously unrecognized predatory instincts. This is, of course, not an expected outcome but a possibility that demands parental awareness. Do not leave any vulnerable child alone with any animal, human or otherwise, whose trustworthiness has not been tested over time.

It is also important to recognize that even gentle and tolerant pets can unintentionally harm a child. A baby can be knocked over and injured by a friendly dog or a cat that affectionately brushes against the infant. A child carrying an ice cream cone may get nipped or scratched accidentally as your pet aims for the treat. Keep the nursery door closed or install a screen door that can be securely closed instead. Netting installed over an infant's crib can afford safe slumber and discourage your cat from cuddling too close. Until your baby has sufficient motor control and strength to move independently, dogs or cats should not be permitted to sleep near their faces. Basic hygiene is always prudent and will prevent the transmission of diseases such as internal parasites.

If kids are fighting, a good dog will jump on them to break it up.

Parents frequently report that the family dog acts as a kind of referee/baby-sitter when their children fight. The truth is that the family pet is most likely responding to the activity in a nonspecific fashion. Dogs are normally attracted by the level of arousal generated by the people around them. The family pet can be expected to jump into the action of a conflict between children just as it would join in their playful antics in the backyard. Dogs are, indeed, astute observers of human body language and tone of voice. They may well sense the tension between people in conflict. This does not mean, however, that they are capable of intentionally separating adversaries. Regardless of why a dog intervenes in an argument, perhaps the only thing that really matters is a peaceful outcome.

If a dog seems harmless to an infant, there should be no problem between them as the child grows.

The vast majority of pet dogs are tolerant and patient with the children of the family. For some dogs, ancestral instincts can emerge with tragic consequences.

Predatory aggression is common to all dog breeds. As I have mentioned, the highest risk period of dog attacks against children occurs in the transitional toddler phase. This sensitive phase requires the family pet to make the connection between the immobile infant and the newly ambulatory baby. A dog with deeply rooted hunting drive may perceive the toddler as an unfamiliar target or strange intruder. Although it is difficult to predict which dog is more likely to make this error, several predisposing trends emerge. Hunting dogs that are prized for their abilities may never hurt a child, but any demonstration of aggression toward small prey must be considered a liability. Dogs that have a history of running away may be less reliable and may have had the opportunity to express predatory instincts that would have been restrained in the backyard.

Dogs that are unresponsive to Obedience training or those that have yet to be trained are less easily controlled in almost any situation. Pets that become aggressive near food or possessive about a favorite bone or toy might not hesitate to punish an approaching toddler, even when the child has no intention of removing the valued object.

Parents should be vigilant over a toddler in the presence of an unrestrained dog. Even when predatory tendencies are not a concern, toddlers can be frightening to the family pet. *In fact, pets are often more afraid of babies than babies are of them.* A small person with chubby fists waddling or crawling at rocket speed can terrify any pet.

In most cases, you will need to protect your dog from the attentions of your baby. However, because the potential injury to your child by your dog is far greater at this stage (whether it is predatory or in self-defense), every precaution should be taken to prevent the opportunity for harm all around.

Supervise your children and pets. Introduce each to the other gradually and with planned and supervised sessions. Give your pet positive reinforcement for desirable behavior when your child is nearby. Give your child encouragement and praise for gentleness toward your pet.

Male dogs are more likely than females to become aggressive toward a new baby.

Aggression toward a newborn baby is more likely in a male dog. The type of aggression most likely to be shown toward infants is probably more closely related to predatory aggression, which is not influenced by sex hormones. Aggression directed toward any new addition to the household is not a function of the dog's gender, but rather its individual instinct and experience. Predatory instincts in the dog may be triggered without prior experience. Male or female, your dog should be gradually introduced to the new baby under controlled circumstances. Newborn babies should never be left unattended with any unfamiliar pet or person.

Male dogs may be less tolerant of young children, whose attraction to pets can occasionally be overly enthusiastic. This does not mean that male dogs are more aggressive with children. Aggression is not the only way for a dog to avoid an unpleasant situation. Most dogs, male or female, will usually try to leave the source of undesirable attentions before they are pushed to the point of becoming aggressive.

Not every female dog is tolerant of children (or anyone, for that matter) that awaken them from a nap in their favorite corner. When dogs cannot escape unwanted attention, however, males may be more likely to show irritable aggression by growling or curling their lip. These signals are meant to warn of their increasing arousal and should be respected. Signs of aggressiveness need not result in acts of aggression if a dog's warning is heeded.

Dominance aggression exists in both males and females but is more common in male dogs. In general, children are less able to command authority and a socially ambitious dog may sense its advantage. A dog may interpret a child's affectionate hug as an attempt to assert dominance. This may cause a dominant dog to retaliate and put the child "back in place."

Young children must be taught how and when to interact with a family pet and with pets belonging to other people. A pet's tolerance of young children is often determined by temperament, life experience and your patient training.

6
Obedience

Obedience training should begin when your puppy is at least six months old or after neutering.

Basic Obedience skills should begin from the moment you acquire your dog. Waiting until your puppy is over six months before teaching any Obedience skills is like a child turning eighteen years of age before going to school! Think of all the delinquent behavior, undisciplined energy, unfocused intellect and utter chaos there would be from such a child! If you wait until your pet is old enough to be neutered to begin setting standards for desirable behavior, bad habits will already be formed.

The average pet is neutered sometime between the age of six months to one year, but it could also be much later. If you acquire an adult, for example, waiting to begin training makes no sense whatsoever. The presence of an intact reproductive tract does not interfere with the brain's capacity for learning!

If a puppy is too young to join a local puppy "kindergarten" class, you can begin teaching basic skills yourself. A pup as young as eight weeks of age can begin to learn "Sit" or name recognition.

Repeat the name until you get the dog's attention and then give enthusiastic verbal praise ("good dog!") for looking at you. As the pup begins to move toward you say "Come" with open arms and a big smile and reward further with a piece of dry dog food or a pat on the head (or a kiss!).

Discourage nipping or mouthing your hands or clothing by directing a pup's need to chew toward an acceptable substitute, such as a toy or chewable rawhide bone. Discourage jumping, even at this young age, because it will persist in the adult dog. In fact, discourage jumping, nipping and anything else that you would not want them to continue as an adult.

Retrieving a ball or chasing a Frisbee instead of a person is much more acceptable. These games, which can continue for a lifetime, apply Obedience

Practice Obedience inside your home and outside, too, where distractions will challenge the reliability of training in any situation.

skills such as "Fetch," "Come" and "Drop it" or "Give" in a setting that is fun for dogs of all ages. Learning Obedience skills in the context of play is perfect for young pups and can be applied later in other circumstances.

Speak with your veterinarian for recommended puppy Obedience classes in your area. Learn the fundamentals of Obedience so that you can learn to shape your puppy into your ideal of a dog. These classes also enable your pup to interact with other dogs to become a more sociable canine citizen. Most puppy classes begin around three or four months of age. Be sure that your puppy has had at least two in the series of inoculations before you begin. This is the time to expose your pet to a variety of experiences (visit your friends, go for car rides, meet children in the park, play with other puppies and friendly adult dogs) so that you can set the groundwork for desirable responses in any situation. Start basic training at home before classes start, and spend as much time with your puppy as you can. Waiting until later is almost always a mistake.

Physical force is the best way to gain a pet's respect.

The use of physical force does not lead to a pet's respect, only temporary control. Restraint is sometimes necessary for a pet's own benefit, such as during a veterinary examination, or for the dog's own safety, by using a short and taut lead while crossing a street in heavy traffic.

Training exercises, such as placing young and cocky dogs in a submissive "roll," can be helpful for dogs in need of a quick attitude adjustment. *Notice, however, that physical force is never physical abuse.* A petite woman (or a man, for that matter) could master the power of a male Rottweiler but will never succeed if she relies on physical strength alone.

Dogs work well if motivated by pack unity or defense. Members of a human family are substitutes for the social group of wolves and dogs known as packs. Continually punishing and physically forcing a dog could weaken the emotional ties that bind the dog to you. The dog's behavior could become inconsistent and based on unstable pack dynamics.

Cats are even less tolerant of punishment issued directly from an owner. The cat's social nature is often more cautious, less obviously social than the dog. The cat is not a pack animal and is not motivated by ancestral instincts to cooperate in groups (with the exception of lion prides). A mistreated cat has no reason to bond to you in any way.

If your intention is to change your pet's behavior, it can be helpful to think in terms of shaping behavior toward a desired response (or gradually away from an unacceptable one, as the case may be). Rather than acting as an abrupt enforcer who demands immediate correction, think in terms of improving your own patience and tolerance. Concentrate your efforts instead toward progressive improvement accompanied by gentle and consistent guidance. Rigid and stern disciplinarians are out of style, and for good reason.

A food reward must always be given during training.

For most living things, food is a strong enticement. For more intelligent and evolved animals, other things besides food can attract the individual's interest almost as much or more, particularly when it is not hungry. Alternative rewards for dogs and cats include positive forms of attention such as a gentle caress or back scratch. Quick to interpret our facial expression and tone of voice, a pet can be rewarded by a happy face and pleasant voice.

Food works, but it does not replace love. Besides, you do not want to condition your pet to obey or cooperate with you only for food. For this reason, it is generally recommended to use food rewards sparingly. Instead

"Honey, 'Sit.' Good dog!"— Obedience skills should be practiced throughout the day and in more formal instructional sessions.

of giving your pet the entire treat, break it into small portions. Better yet, do not use pet treats at all. A single tidbit of dry pet food is sufficient. Some dogs will do anything for an ice cube. If you give food as reward, it is probably better not to use only food. Instead, give food rewards only intermittently, every other time and then eventually at unpredictable times. Alternate food rewards with a kind word or a scratch behind the ear. Kisses are good, too.

Punishment must be painful or cause fear to be effective.

The selective use of controlled punishment has a very limited place in pet training. Punishment may be loosely defined as the association of a negative experience with an individual's behavior in view of decreasing the reoccurrence of that behavior.

To be effective, several important qualifying elements regarding the use of punishment must be considered. First, the behavior being punished should be really worth punishing. It might be something dangerous, like chewing on electrical cords. Second, the choice of punishment should be viewed as a negative experience by the offender. Treating the surface of a chair leg with Tabasco sauce will not deter gnawing on it if a dog likes the taste of Tabasco. Third, the punishment really should fit the crime. Abuse, whether it is psychological or physical, is never advised. It is unwarranted to socially isolate a young puppy in a crate for urinating in the house. *It is never appropriate to beat or kick a pet.* Finally, the negative experience must coincide with the undesired behavior. It is useless, for example, to punish a cat for scratch marks on the sofa when you have no idea when the scratching occurred.

Still, punishment can have a place. The choke collar is a common form of discipline. When tension is applied with a quick jerk and immediately released, a dog's hesitation or reluctance to obey is discouraged. Choke collars with blunted prongs apply a pinch as well as a hold on a dog's neck. These collars should be reserved for dogs that are slow to progress with necessary Obedience skills and only when their neck muscles are well developed. Excessive use of this punishment will injure a dog's neck, including the cervical vertebra and the trachea, or windpipe.

For small breed dogs, harnesses may be best, but serve only to restrain and not really to train. Halter-type collars provide a nylon choke with an additional loop around the muzzle. With gentle tension, this collar brings the dog's head into a submissive downward position with little effort on your part. Halter-type collars work well for dogs of any size, but are particularly useful for small ones that are difficult to correct without injuring your back!

Shock collars, in contrast, are a perfect example of inappropriate punishment in virtually all circumstances. Collars that apply an electric shock to a dog's throat are advocated by some dog trainers. These collars also form the basis of the electrified underground fences. Shock collars are almost never warranted, especially when other more humane approaches to training have been overlooked. Try one on and see for yourself. Collars that employ ultrasonic signals instead are generally ineffective but at least they do not harm the dog.

Other forms of acceptable discipline include the "alpha roll" in which a person gently but firmly rolls a dog onto its side or back in a submissive position (submissive roll) and prevents it from standing. This can be effective in puppies that challenge their owners with shows of dominant bravado. The "roll" should be used with extreme caution or not at all in adult dogs that have already proven their aggressive tendencies. Nipping or inappropriate mouthing in young dogs can be discouraged by enclosing the muzzle in your hand and applying momentary pressure. This requires frequent repetition but can be very effective. Spraying a cat with water can keep it away from investigating counter tops or chewing on plants. The advantage of this technique is that the water bottle, not you, becomes associated with the punishment. Some cats like water, however, or endure the punishment if they really want to persist in what they set out to do.

Punishment or discipline is more effective than reward in modifying undesirable behavior.

It is in the nature of most normal living things to prefer pleasure to pain. Let us define punishment as the association of a negative outcome with an action that will decrease the likelihood of repeating the action in the future. Pet owners should not need to resort to corporal punishment to correct undesirable behavior. It is important to realize that animal abuse need not be as extreme as beating a dog with a stick. Discipline becomes abusive punishment when it exceeds what is necessary to discourage a pet's undesirable behavior.

A dog will work to earn the praise and attention of a loving person. A dog will become neurotic and withdrawn in misguided or malicious hands. Although mild forms of discipline can be effective teaching tools, many are often misused and overused.

One of the major problems with relying on discipline/punishment alone is that, although it may temporarily stop an unwanted behavior, it does not teach your pet a desirable alternative activity. If discipline is not immediately followed by teaching what the dog could be or should be doing to earn your praise, a valuable learning opportunity is lost. You also lose your time and energy in unconstructive and negative emotion. You lose your own self-respect. You could lose your pet's trust and sense of security for fear of physical punishment when approaching you. Punishment creates an air of tension and mistrust, which interferes with learning anything other than fear and anxiety. Ultimately, punishment is not conducive to learning. Reward-based learning, or positive reinforcement, is usually more constructive in the long run and is easier on both owner and pet.

If you continually feel frustration, helplessness and even rage toward your pet, you are doing something wrong. If you feel a need to resort to hitting or socially isolating your pet to gain control, you have already lost, and you need help. If your veterinarian cannot advise you, ask for the name of a good Obedience class or a veterinary behaviorist who can give you private evaluation and guidance tailored to your particular needs. Most big problems begin as little ones. Little problems typically require little time or effort to resolve.

Striking your pet with a rolled up newspaper is effective discipline.

To effectively discourage the repetition of a problem behavior, discipline should be appropriate and immediate. Striking at or near your pet with or without an object as a weapon is inappropriate punishment. By the time you find a newspaper, roll it up and chase your misfortunate victim, it will be too late. Your pet will no longer be able to make any clear association between its action and your reaction.

Striking your pet with any object is never acceptable. The only thing you will teach your pet by brandishing a newspaper or magazine is to develop a fear of newspapers, magazines and you. Fear is not respect. Fear is not love.

Hitting at or near your pet with a rolled up newspaper is a useless training method with the only likely result a fear of newspaper and you.

Instead of using it to threaten your pet, recycle your newspaper instead! Relying on punishment alone to correct a problem behavior is a lost opportunity to teach a desirable alternative.

In almost every situation, it is better to train your pet by positive example and to prevent the need to repeat an undesirable behavior. If, for instance, your kitten knocks over a plant, remove the plant from the kitten's reach and say "No." Then, give an attractive toy or spend some time in interactive play. If your dog chews on your shoes, say "No" firmly and remove the shoe. Immediately give your dog a rawhide chew or other chew toy and say "good dog!" Increase your pet's activities in general, and as your mother used to plead, put your shoes away!

In most cases, the best way to punish your pet is not to punish at all. It is always better to emphasize the positive by using reward for good behavior. Whenever possible, direct your pet toward desirable behaviors and provide plenty of opportunity to perform these. Immediate and abundant praise will present a clear association in your pet's mind on how to please you. Give your pet praise for resting quietly. Take your dog for a walk before barking to gain your attention. Should undesirable behaviors occur, try to be patient. Concentrate instead on understanding what circumstances led to the display, and work on preventing the same circumstances from being repeated.

Obedience training is unpleasant for dogs.

Obedience training teaches your dog to associate key words or phrases with a specific behavior. In a sense, it provides a primitive kind of verbal communication. Applying the commands, the dedicated pet owner can encourage the dog to behave in desirable ways in almost any situation. Successful performance of a required task must be closely followed by abundant and exuberant praise. To teach these skills, good dog owners know that they must invest a lot of time and patience. To learn the basics, a dog needs its owner's uninterrupted attention on a daily basis. Practice these skills everywhere inside the home and during walks outside, too. Despite increasing distractions, the dog's obedience should be expected and reliable everywhere. It also increases the benefit, both physically and intellectually, of every walk.

Obedience skills can and should be practiced spontaneously throughout the day, in more formal training sessions and even during goal-oriented play. Games like fetch incorporate the "Come" command as objects are retrieved to you. To earn back the toy, your dog must "Sit" or "Down." Now, does any of this sound like cruel and unusual punishment? Better than discipline, get up and try playing fetch with your dog.

Small dogs do not require Obedience training.

A Chihuahua can have the heart of a lion and a Great Dane can be as timid as a field mouse. Lhasa Apsos can be dominating and fearless while German Shepherd Dogs can be submissive and afraid of their own shadows.

Dogs have no real conception of their size. They learn about their abilities and limits by feedback from the world around them. Owners of a small dog might feel obliged to protect their pint-sized pup from potential mishaps, but should beware of overindulging. What the pet owner perceives to be devotion and protection might convey entirely different messages to the diminutive dog. A dog that is encouraged to sit in your lap or be carried around is continually in a physically superior position to you. This is one of the ways that a dog learns that he or she is, quite literally, at the top.

A little dog that is allowed to jump on people, even in apparently friendly greeting, or jump into their laps, especially uninvited, is being rewarded for dominance-related behavior. A Toy breed dog that growls and barks at visitors could be whisked into the owners arms to be cuddled and caressed. Thinking that they are comforting a frightened creature, concerned owners might not realize that the dog perceives this as reward for territorial aggression. Many small dogs are, indeed, frightened in unfamiliar situations, but they are not born anxious and afraid. Nervous and defensive dogs are often created, mostly because they are sheltered and rarely exposed to new people and places. They tend to be walked less often and more briefly compared to larger breeds, partly because owners worry about their resilience to weather. Aside from a possibly greater need for sweaters and coats when the temperature drops, a small dog is essentially no different from a big one.

Obedience training actually decreases a dog's social insecurities because you are able to communicate exactly what is expected of the dog in any situation. Part of a dog's social anxiety, for example, can stem from not knowing how it is expected to behave. Instead of becoming anxious and unfocused, your dog can be directed to a "Sit/Stay." By giving a command, you relieve your pet of decisions. If you train your small dog as you would a large one, you will have a well-adjusted and even-tempered companion. If you allow bad habits to persist uncorrected and do not instill your pet with limits and standards of desired behavior, it will be a source of ongoing disappointment to you. Size has nothing to do with that.

A good watchdog or hunting dog must be kept outside at all times

(See chapter 3, "General Misbeliefs About Dogs," section entitled *Dogs are meant to live outside in cold weather.*)

Deaf dogs cannot be Obedience trained.

As long as their vision is not affected too, hearing-impaired dogs can be trained to respond to hand signals.

There are many conditions that cause deafness in pets. Just like people, viral and bacterial infections can destroy the nerves responsible for sound detection. Certain breeds of dog have a higher incidence of congenital deafness. The Dalmatian, Samoyed, Old English Sheepdog and Australian Shepherd are only a few of the breeds affected. Deafness can be bilateral and profound, or it may affect primarily one ear, or extend over a limited range of sound. Consult a well-reputed dog trainer in your area who also has experience in training with hand signals. A hearing-impaired dog can compensate beautifully by relying on vision, vibration and odor. It may present additional challenges, but your efforts will certainly be worthwhile.

Cats cannot be trained to do tricks or respond to commands.

Most people do not expect from a cat what they would from a dog. There are, indeed, many species-related differences, yet individual cats and dogs show a wide range of temperaments and talents. Not all dogs can be trained to retrieve. Some dogs, such as Chow Chows and Akitas, can be particularly resistant to obeying commands. In some ways, the nature of these breeds resembles the feline stereotype, cooperating when it suits them but adoring their owner nonetheless. Some dogs are more like cats, and some cats are more like dogs.

Cats can be taught to retrieve objects and respond to other commands as well, if you take the time to train them. Note that this cat is carrying a toy.

If you never try to teach any Obedience skills or tricks, your cat probably won't learn any. Occasionally, cats that are socially interactive and intelligent still manage to find clever ways to attract your attention and seem to spontaneously develop what a dog might take weeks to learn. You cannot say cats cannot be trained without trying to find if there is something unique that will gain your pet's interest.

My first cat, Jonathan, used to love to go for walks. Cautious at first, he became accustomed to a nylon harness and leash and seemed to walk almost proudly with his black bushy tail erect. He became so comfortable walking outside that we began to jog together. He kept pace, trotting happily (and effortlessly, I might add with envy) beside me. Another favorite pastime was a game of "fetch" in which he chased a ball at top speed and dropped it at my feet. A sort of ping-pong game was also fun in which we would push a ball back and forth to each other. Although tragically short-lived, he was a special kind of cat, with distinctly canine qualities.

His companion cat, Sara, is very dainty but reserved and intermittently condescending. A fifteen-year-old princess (she prefers not to discuss her age), she will respond to "Come," but only if she is hungry or not napping. Her current housemate is Hershel Walker, affectionately known as Hershey, a two-year-old chocolate tabby who is quite brilliant. Because I no longer jog, having returned to the study of ballet (the other inspiration for his name besides the coat color), we have found other diversions. Hershey, too, tirelessly retrieves favorite toys. Among these is a long plastic stick, all that remains of a toy that used to have long feathers at one end, which he carries in his mouth and intentionally sticks into objects and under doors. When he is done sticking the stick, he retrieves it to begin the game all over again. Hershey consistently responds to "Come" and "Sit."

You, too, can challenge your cat's hidden abilities and tap into undeclared interests to reveal the dog in your cat. Of course, a cat that is a cat is still special, too. Just ask Sara.

The five basic Obedience commands are "Sit down," "Lay down," "No," "Come here" and "Get down."

The five basic commands are "Sit," "Down," "Come," "Heel" and "Stay."

Each word is used consistently in the *same form* so as not to confuse the dog. This strengthens the link between your verbal command and the pet's expected response. "Sit" should always be used so that "Sit down" remains distinct from "Down." The simpler the command word, the easier the dog will learn. There is nothing wrong with saying "Lay down" if it is used consistently, but it is more efficient to simply say "Down." Similarly, "Come" should be the habitual command so that inconsistent phrases such as "Come over here," "Come on" and "Hurry up and come here" are eliminated.

Eventually, when the five basic command words are recognized and obeyed, you can build upon them and introduce additional commands that might contain more than one word (as long as they are used in the same form each time). Begin each command with your pet's name in a crisp but kind voice, like a snap of your fingers. You need the dog's (or cat's) attention to give an instruction. It makes no sense to say "Sit-Stay-Sheba" if the dog is distracted. Say "Sheba Stay," then "Good girl!" and you will not be wasting your time or breath.

Words like "No" are not commands at all in any practical sense. What "No" means to you and to your pet is that you are not pleased with what is currently happening. Thus, "No" becomes a mild form of punishment. "No" can be an effective way to startle your dog because activity may be momentarily interrupted so the dog will give you attention. Used alone, however, startle commands do little else. Your pet may stop briefly. But unless you use this window of opportunity to direct the dog toward a more appropriate form of activity, your chance to teach something really valuable is lost. If your pup is chewing on a shoe, for instance, say "No" as you remove the shoe and immediately replace it with a chew toy and say "Good dog!" This clearly demonstrates what the pup should not be doing (how *not* to earn your praise), and what would be a good substitute (how to earn your praise).

Phrases such as "Get down" are commonly used, for example, by frustrated owners to keep agitated dogs from jumping on people. In this context, "Get down" is no different from "No." Unless the dog's action is directed toward a desirable alternative, the dog remains in control over what it chooses to do next. It can either continue to jump on people or find some other distraction that you might like even less. Instead, use it as an opportunity to teach your dog to greet people in a calm and submissive way by commanding "Sit" and "Stay."

In general, verbal praise should be given in a calming and soothing tone. This becomes important, for example, if you are praising your dog for a "Sit." If your dog is in a calm and controlled position and your praise is given in an excited and loud voice, most dogs will respond to your excitement and break out of the "Sit."

On the other hand, the "Come" command is the exception. As your pet approaches you, be as enthusiastic and joyous as you can be. Why else would it approach you? The "Come" command should be said and praised with a happy and light tone of voice. Never call your pet to "Come" if you are angry or anxious. It will quickly learn that "Come" means to run away.

The other commands should be firm but no less worthy of soothing forms of your praise. Always give your pet praise, in the form of encouraging words or a small food treat or a caress. That is what reinforces desirable behavior and ensures that the behavior will be repeated.

Never deprive your pet of your praise, even long after an obedient response. Praise is what your pet is working for, it is what we all thrive on. To deny someone you love, and that so obviously loves you, the pleasure of your approval and appreciation is no different than skipping a meal or removing drinking water or ignoring. Your praise needs to be earned, but it should be liberally given. Praise must also coincide closely with good behavior to be most effective. Delayed gratification does not work with pets (it barely works for most people!). Immediate and abundant praise makes it very clear to your pet exactly what to do to earn it again.

7
Aggression

Do not disturb a sleeping dog.

Imagine being suddenly roused from sleep by a small child screaming in your ear or by anyone's unexpected touch. It is normal for sleepy individuals to overreact to situations which are potentially threatening. Indeed, an immediate defensive response in the first few moments of consciousness can decide an individual's survival. In vulnerable contexts, instinct directs us to act first and ask questions later.

Sudden disturbance during a moment of inattention occasionally results in a pet's aggression. This type of aggression is likely rooted in fear but may also contain elements of irritable aggression. (In people, this type of aggression is often referred to as irritability, crankiness or annoyance). Unexpected attention may also include a special type of territorial behavior. Territorial defense of a favorite resting place may be triggered by unwelcome intrusion and could amplify a pet's underlying fear and/or irritability.

If your pet is intolerant to interruption from a nap, give some advance warning. Call the dog or cat's name gently but loudly enough to waken the animal. Before approaching any further, make sure your pet is clearly aware of your presence. Another option, particularly if your pet is upset, is to call the animal toward you and away from the resting place. This is a simple way to diffuse the situation and avoid a predictable conflict. No one enjoys being shocked out of peaceful sleep, and we cannot expect our pets to be more tolerant than we are.

Dogs will never bite the hand that feeds them.

Canine aggression is directed toward familiar and loving hands every day. Dogs behave aggressively for many reasons. Given the right set of circumstances, at least in dogs' minds, they may very well bite their owners or anyone else.

An animal in deep pain, for example, can strike out at anyone that approaches, even in an attempt to offer aid (irritable and fear-induced aggressions). Dogs that are aggressively possessive over an object they consider valuable, such as a bowl of food or favorite chew toy, will actively guard it from anyone that threatens to remove it (possessive aggression). A cornered and frightened pet will do whatever it takes to defend or to ensure a quick getaway (fear-induced aggression), even biting the owner. A dog that has become socially dominant over an owner may retaliate if the owner, even unknowingly, commits an act that challenges the dog's authority (dominance aggression). Aggression, in many forms evolved as an essential element for existence and survival. Indeed, the history of human civilization is built on "justified" acts of aggression.

On the other hand, there are countless examples of dogs that could bite, indeed probably should have bitten, in many adverse situations but did not. The inhibition of aggression is under complex mechanisms of central nervous system control. Severely injured dogs will often lie quietly as the veterinarian treats them. Extremely frightened animals can be passive and freeze rather than become aggressive. The nobility of the dog is not diminished by the ability to show aggression.

Female dogs make better watchdogs.

Territorial aggression is very much an individual quality among dogs. There is no proof that females are superior guard dogs over males. There is no evidence that males offer better defense than females. In fact, the statistics indicate that a dog's gender is of no significance in determining value as a watchdog.

Some studies suggest that the breed might be a factor in a dog's guarding abilities, but there is so much variation within a breed that this too should be considered of secondary importance. The best natural watchdogs include the smaller breeds that may be more easily startled, quicker to bark and completely oblivious to their size.

The average dog is easily trained to become what the average pet owner would call a "good watchdog." Reward your pet for a startled "Woof!" when something is heard outside or if someone approaches your door. Your reward may be in the form of verbal praise ("Good dog!") or simply your attention (such as saying, "What is it, Max?" or going to the window to see to what your dog is reacting). Of course, negative attention is still attention that might reinforce a pet's behavior. Even if you instruct your barking dog to 'shut up', any form of your attention can be reinforcement for territorial behavior. If you think that you need a dog prepared to attack an intruder, contact a well-reputed dog trainer in your area for specialized training.

The encouragement of uninhibited aggression in a pet dog should be left to the skilled professional, and you must become equally skilled in controlling the dog. Dogs that are attack-trained are not the ideal family pet for most households.

Play-biting is harmless and pups soon outgrow it.

Playful behavior is not just random chaos. It is Nature's way of allowing young animals to learn about themselves and the world around them in a relatively safe way. Juveniles can practice all sorts of physical maneuvers and contortions to improve agility and strength. These will become invaluable skills in adulthood. Playtime is serious business. Play incorporates sequences of behavior related to important social interaction, escape from danger, self-defense, hunting and even mating. Young animals learn about their own abilities and limitations in a context with few serious consequences. They learn what works to influence others to get what they want or to prevent what they do not want. Play-biting is a very effective way to accomplish many things.

For most pet owners, an adult pet that bites is almost never acceptable. Pups or kittens that are permitted to practice undesirable behaviors during play are much more likely to persist in these patterns as adults. For this

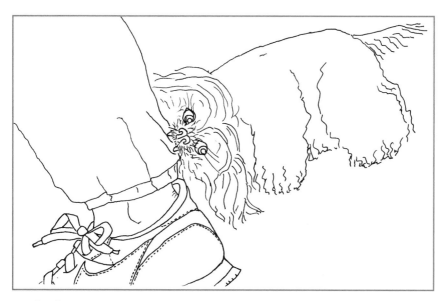

Play-biting in a young pet should be immediately discouraged if it is not to become a part of adult behavior.

reason, play-biting directed toward a person's clothing or any body part is not cute or harmless at any age.

Games that encourage a puppy or kitten to chase children around the yard, or pounce at wriggling fingers or toes, or nip at someone's heels as they pass should be strictly discouraged. If your pet becomes wild, agitated and difficult to control when it plays specific games with you, choose different interactions next time. Games which divert energies toward moving objects, such as "fetch," are no less fun and far more appropriate in the long run. If you are unsure as to whether a game is appropriate or not for your growing pet, ask yourself whether the behaviors associated with that game will be acceptable when it reaches adulthood. Your answer will be plain to see.

Puppies have to learn that there are two sets of behavior. What is permissible behavior with other dogs may not be okay with human beings. They must be taught that there are limits for their interactions with people. It is up to you to set and to teach these standards of behavior. It is normal for puppies to nip, but it is the wise pet owner that recognizes the implications and consequences. Respond by a gentle but firm "no!" and direct attention toward an appropriate chew toy or ball. If the nipping persists, gently grasp the pup's muzzle and hold it closed while you repeat your warning "no!" Then repeat your diversion toward a desirable alternative activity. Puppies can be quite persistent at nipping, but you must be ready for it and answer them with consistent and quick discouragement. Eventually, they will understand that playtime has its limits, too.

A dog should be encouraged to play rough and wild games in order to behave more calmly later.

Many people encourage their pets to play games like "tug of war," or tease them in wrestling matches or games of chasing each other. This is one of the most common errors of raising a dog. Games like wrestling, chasing people and "tug of war" teach dogs that wild and uncontrolled behavior targeted toward people is not only acceptable but desirable as well. Particularly for young dogs, this can have lasting consequences in the appearance of undesirable behavior in adult life.

Puppies, for example, that are teased with tugging on objects frequently growl and become agitated. This game teaches pups how strong they are, physically and socially, in relationship to playmates. Even if the puppy is allowed to "win" only occasionally, it will be promoting an attitude that most dog owners will find unfavorable. The function of play is not only to provide an outlet for energy but an opportunity for learning. Play is a valuable teaching tool that provides the owner with the opportunity to shape the dog's behavior in general. Most pet owners do not want a dog that bites, jumps on people, chases children or rips at clothing. Yet these are behaviors encouraged in inappropriate forms of play.

Avoid playing games that seem to make your dog uncontrollable or aggressive in any way. Rather than encouraging chaotic and random entertainment, play time should direct your dog's energies toward constructive games.

Practice Obedience skills as a form of play. Get your dog involved in Agility training through a local dog trainer or kennel club. Agility training incorporates Obedience skills with activities like climbing inclined boards or jumping over obstacles. These are challenging to both pet and owner. Play games that direct the dog to chase or bite neutral objects like a rubber ring or ball. Give "Sit" and "Stay" commands before throwing the ball. Use "Come" and "Sit" to teach the retrieve of any object. Use an additional command, such as "Drop it" as you gently pry the object from your dog's mouth so that it eventually will be relinquished voluntarily. All these skills are no less exciting to the dog and will be of real benefit to a lifetime of contented canine companionship!

Dogs are always aggressive toward postal carriers.

Pity the poor postal carrier that braves rain, snow and sleet only to confront a defiant and snarling dog. Not all dogs, however, are born with equal determination to defend their territory against the daily intrusion of mail. Aggression against postal carriers can originate as the act of a fearful dog startled by the clank of the mail box or the thud of dropping mail into a peaceful home.

Whether their attitude begins out of territorial prowess or self-defense, many dogs seem to eagerly anticipate the mail delivery as if it were a peak form of entertainment. For many dogs, the arrival of the postal carrier is truly a high point of their day. Dogs that have few outlets for exercise and intellectual stimulation will be more likely to overreact to events throughout the day. Dogs that have had a good long walk in the company of their owners, some play time and a fun session of reviewing Obedience skills, might even sleep through the mail delivery!

If your pet is a menace to your postal carrier, you have two choices. You must ensure that your dog is securely confined so that the delivery can be made in safety. Or, if the carrier is willing, arrange for them to meet on neutral ground. Have the carrier offer your pet a treat. Practice this over many days and at a distance closer to your home and front door. Practice inside your front yard. Your postal carrier might agree to come in after work dressed in street clothes. If the person is recognized as a friendly intruder your dog

Games like tug of war are one way that dogs learn who is the strongest and most determined, helping to clarify dominant and submissive ranks. (photo courtesy of Estelle Schwartz)

will be less easily triggered by the sight of a uniform that once was so strongly associated with a negative response. Leave a small supply of dog treats or dry dog food in your mailbox. Ask the postal carrier to include a treat with your letters or toss your pet a snack. Your dog will look forward to the mail delivery and so will your postal carrier!

Mounting behavior indicates a dog is sexually aroused.

One of the more misunderstood and peculiar dog behaviors common in both males and females is "mounting." Mounting is a somewhat comical exhibition with very serious social implication. It occurs as a dog clasps an object or individual with its forelegs and mimics breeding motions.

Mounting can be socially motivated as an extreme form of dominance-related aggression. It can be directed toward other dogs but is frequently practiced on children and adults, too. If your dog is "humping" your leg, or your child, consider these actions to be politically charged. Tolerance of mounting, even by young puppies, provides the dog with the unintended feedback of having social prominence over a person. This should be firmly discouraged. Mounting is not only an expression of the mounting dog's social nature, but also can be sexually motivated. Many behaviors serve more than one behavioral function and can be displayed for more than one reason. To interpret this behavior the observer should identify the dog's target and interpret the situation.

Dominance between dogs is frequently declared or challenged by relatively inconspicuous behaviors which may include elements of mounting behavior. Mounting behavior may not seem so inconspicuous, but it is subtle compared to dogfights that erupt between dogs battling for superior social status. Mounting to establish dominance between individuals can be exhibited by and between individuals of either gender. Mounting may be seen in male dogs that are neutered too, especially if they had sexual experience prior to being castrated and if there is a female "in heat." Sexually motivated mounting can be directed at neutered female dogs but is clearly more intense if the bitch is "in heat."

It is never appropriate for family pets to feel socially superior to people. Tolerance of mounting directed toward people is the same as encouraging the dog. The mounting dog should be gently but firmly placed in a controlled "Sit" or "Down" and "Stay." Be aware of how your position facilitates this behavior. Discourage people from sitting on the floor with the dog. Instead, keep the dog at your feet and off the furniture. You can enjoy your pet's company and still maintain your social dominance if you remain aware of subtle social manipulation.

Mounting during play is one way that social dominance is established without fighting.

Mounting behavior is not seen in neutered dogs.

In neutered males and females mounting behavior is far more likely to be a declaration or challenge of social rank than an indication of sexual attraction. Neutering effectively eliminates sexual hormones from a dog's system. Mounting behavior, however, has important social function independent of sexual motivation.

The presence or absence of reproductive organs and hormones has little impact on a dog's social rank. The only thing these hormones may be said to influence is the intensity with which an animal will pursue social ambitions. Neutering will tone down the intense rivalry for social dominance between dogs, but does not alter a dog's basic character. Aggression will be somewhat subdued after neutering, but often persists even in neutered males. It can be difficult, if not impossible, to distinguish from dominance-related conflicts. Sex-related mounting is usually eliminated by neutering. A male with prior mating experience may attempt to mate with a female in heat, but such attempts will be, quite literally, fruitless, as neutered males are capable of erection but sperm production is impossible.

Dominance aggression is always expressed by mounting.

Mounting behavior can be, and most often is, quite subtle. The dominant dog can exhibit more subtle elements of the behavioral sequence culminating without actually mounting. The dominant, or "top dog," can simply place a head or paw over the other's neck or shoulders. The dog that is on top is asserting dominance over the other. The dog under the nose or head or paw or body of another dog is being given a blatant message that they are, indeed, a socially inferior "underdog." The "underdog" can either accept this declaration of lower status or take up the challenge. Subtle dominance behaviors have similar social impact because they are isolated elements of the more obvious mounting display.

Mounting behavior in female dogs indicates sexual confusion.

Mounting, when unrelated to reproduction, has a social significance that applies to both males and females.

Mounting behavior, as a display to challenge or to assert social domination by one dog over another, can occur between individuals of the same or opposite sex. Females can be dominant over other males and other females. The dynamics of dog society is more a question of physical ability, combined with temperament and drive, than a question of gender alone. In a group of dogs as in their wild relatives, the wolves, a lead female emerges from the ranks, as does a lead male.

Mounting behavior does not signify that a bitch is confused with regard to her gender. Rather, mounting by a dog, regardless of gender, is a very clear signal of serious social ambitions. Mounting by a female, in a nonsexual context and directed toward another dog or a person's leg, is not the act of a confused dog by any means.

Mounting behavior in dogs begins at puberty (about six months of age).

Mounting can be seen in very young puppies soon after they begin walking and playing with each other. The behavior is a socially significant one in the scheme of canine social behavior and eventually incorporates the people that become familiar as the pup matures.

To young puppies, mounting is among the earliest opportunities for feedback regarding their physical abilities and social potential. This also provides a basis for their earliest attitudes toward other dogs and people. Their physical advantages or disadvantages will be intimately influenced by temperament and such qualities as stubbornness and assertiveness, cautiousness and submissiveness, eagerness to please and social moderation. Mounting by young puppies in a litter is the introduction to many types of physical and social skills that they will continue to practice as they develop. Each pup's behavior will be further shaped by the experiences gathered in many contexts while continuing to investigate the world and to learn responses to self-exhibited behavior. Testing their physical strength against each other, they will eventually apply the social feedback and skill to determine their social rank as adults.

Staring prevents an aggressively aroused dog from harming you.

Direct eye contact with an aggressive animal is definitely something to avoid. Sustained staring is a threat display in many species and the dog is no exception. If the animal is undecided as to whether to launch an attack or not, staring directly into its eyes will likely instigate an offensive.

To the dominantly aggressive dog, a direct and sustained stare is a challenge to social prominence. Dominant animals usually retaliate against potential rivals to defend their social standing. Direct eye contact is also provocative to dogs that are not motivated by social position. This might

Approach every unfamiliar dog with caution, even when the owner is present, and never assume dogs are friendly just because of the breed.

further intimidate a fearful and submissive dog and trigger a defensively aggressive outburst. Similarly, a dog on territorial patrol will be aggravated to attack an intruder that poses the additional insult of staring. Direct eye contact is one of the best ways to provoke many types of aggressive behavior. Regardless of the dog's underlying motivation, the intruder is at great risk.

Joggers or bicyclists are particularly at risk from neighborhood off-leash and unsupervised dogs. Dogs can become territorially defensive to strangers, particularly joggers or bicyclists whose rapid pace is more menacing and gives the dog less time to assess the risk.

Should you encounter a dog that is standing relatively still or approaching cautiously, with eyes fixed on you, the best thing for you to do is to freeze. This is not the time to impress an aggressively aroused dog with your lack of fear or a show of strength. Ignore a wagging tail or position of the ears. These can be misleading and show the dog's internal conflict with regard to the next move. If you make the wrong one, you could regret it.

If a strange dog hesitates to approach you, seems anxious or unsure, or if growling or lip curling are observed even for an instant, this is not a good time to make friends. Avert your gaze to a point nearby which still allows you to keep the dog in view. Back away slowly until you see the dog begin to relax or lose interest. Keep your eyes on the dog until you are safely out of range and consider changing your exercise route. If you feel comfortable, contact the animal's owners when you are calmer, or report your concerns to a local animal control official who can discuss the legal and public safety issues with them.

A dog with a wagging tail is friendly and will not bite.

To understand the body language of dogs and cats, it is important to look at the individual's entire body. Do not focus on the signals of only one part. If you focus your attention at the tail, for example, you will miss information available in the dog's facial expression, the position of the ears and lips, or body posture.

An individual can frequently experience more than one emotion or motivation at the same time. A dog may approach you with a wagging tail, but may also exhibit a curling lip and growling. This apparent conflict in verbal and nonverbal communication is an indication of the animal's own indecision. It is always better to proceed with caution in such cases. Do not rely on the apparently friendly gesture of a wagging tail. It may only serve to distract you from more important messages pertaining to your safety.

Dogs always growl before they bite.

The study of behavior depends primarily on two individuals, the one being observed and the one doing the observing.

Some dogs can be so subtle in the warning that, prior to an aggressive outburst, only the most experienced and alert observer will notice in time to take defensive action. Growling may be so low or so brief that it may be undetected because of surrounding noises or distractions. Sometimes growling may be omitted altogether. Physical signs of impending aggression include a slight curl of the lip or wrinkling nose. This should not be mistaken as a friendly smile. Dogs can curl the lip to expose their teeth during moments of excitement or playful arousal but, at the very least, the lip curl should be viewed as a sign of conflicting emotions in an agitated dog.

When observed in an unfamiliar dog, it should always be interpreted as a warning not to approach. Tail-wagging by a growling dog suggests that the dog is unsure of how to interpret the situation and is in internal conflict as to how to behave. An anxious and insecure dog will bite you even if the tail distracts you from other more ominous warnings. Always err on the side of caution and protect yourself. Any tail-wagging and growling dog would do the same.

All American Pit Bull Terriers are extremely aggressive.

Our ancestors domesticated the dog from a wolf-like ancestor many thousands of years ago. The selective breeding of dogs continued in the process of domestication which emphasizes particular physical attributes and behavioral predispositions. Each breed of dog has become specialized to serve human beings in specific ways. The Border Collie was bred to herd sheep. The Bloodhound was bred to apply a very keen sense of smell. Not every individual of a given breed is born with equal ability to express the desired purpose of the breed. Not every Border Collie instinctively responds to chasing stray members of a flock and may improve little after training. Not every Bloodhound has the stamina or drive to perform a required task.

The Pit Bull, a breed which recently has become fashionable, has a compact and muscular build with a powerful jaw. A characteristic behavioral pattern of this breed is to persist in intense aggression once triggered.

It was originally bred for the entertainment of people who enjoyed watching uninhibited aggression between dogs. This "sport" continues in largely clandestine operations today. Dogs intended to be canine gladiators undergo severe and inhumane "training" to create the intense and uninhibited viciousness desired by their owners.

The behavior of an individual is based on inherited predisposition and the influence of learned experiences. Not every Pit Bull will become instinctively aggressive toward other dogs, although some do. It remains true that individuals of this breed rank high, and out of proportion to their actual numbers, in national statistics of dog attacks against people. Still, the greatest number of attacks come from German Shepherd Dogs.

Any dog that is intentionally trained to attack or, conversely, that is raised with little social interaction and without Obedience training may result in the emergence of latent aggression. Given the predisposition and the opportunity, a dog of any breed can revert to ancestral patterns of behavior that are no longer desirable. Although this breed is not a first-choice family pet, there is no guarantee that a dog of any other breed will not become aggressive toward other pets or people in the right circumstance.

All Golden Retrievers are gentle and trustworthy.

It can be just as misleading to expect all good qualities from a breed as it is to anticipate undesirable ones. In fact, it may be more dangerous not to prepare for potentially aggressive patterns.

Some of the most lethal attacks come from thought-to-be-harmless pets in seemingly innocent circumstances. Just as a Golden Retrievers can have many different shades of golden blond, each must be considered as an individual with a unique appearance and temperament.

In many cases, aggressive outbursts in a dog of any breed are completely predictable in hindsight. They may stem from years of undesirable dog behavior by naive owners.

First-time dog owners must educate themselves by enrolling in Obedience classes. Dog owners are encouraged to enroll young pups from the moment they are acquired in puppy Obedience classes. Owners of adult dogs might benefit from refresher courses, and all dog owners should practice Obedience basics daily, to make sure that their dog's skills (and theirs) remain sharp.

Above all, do not allow your dog to be "trained" by someone else and then returned to you when training is "complete." The purpose of Obedience training is to teach the owner to effectively communicate with the dog, and thereby train the dog to respond appropriately. Your pet might respond to a dog trainer, but what counts is how well you control your own dog.

A pet that becomes aggressive or excessively fearful at the veterinary clinic must have been abused there (or at another clinic) in the past.

For our pets, the veterinary visit can be terrifying, particularly when they cannot comprehend that it is in their own best interest.

When you think about it, it is really quite exceptional that more of our pets do not become uncooperative at the veterinary clinic. Transported in a vehicle away from the security of home, the already terrified pet arrives at an unfamiliar location filled with equally unfamiliar people and pets. Escape is impossible and fear rises. Secure in a carrier, the animal will be further traumatized by being removed and placed on a cold table with no place to hide. All this culminates in being manipulated, poked and prodded by strangers in body parts not explored at any other time.

Fear is a vital survival mechanism. In an unfamiliar setting, it can be an advantage to remain hypervigilant and overreactive. Better safe than sorry.

Fear is the instinctive response of an individual to any circumstance perceived to be unsafe.

The sensation of fear is due to a complex physiological defense system that accelerates heart rate, breathing and awareness in preparation of escape. This occurs along a continuum, ranging from mild anxiety to extreme phobic reactions. Many pets become afraid of the veterinary clinic. After just one visit, the memory of an injection or any other uncomfortable or even painful procedure creates a clearly negative association. Anything that resembles a trip to the vet can trigger the fearful association that was formed even after just one visit.

Occasionally, pets can develop a phobia about the veterinarian. This excessive fear response far surpasses the actual danger of the situation. The anticipation may be so extreme, however, that the individual loses all perspective on the situation and responds panicked. Frightened animals must not be allowed to injure themselves or those trying to help them. Your veterinarian and technical staff may need to hold your pet in positions that seem forceful, or they may require that your pet be muzzled.

Some pets seem less anxious when separated from their owners. If you are asked to wait in another room during a veterinary consultation, rely on your veterinarian's judgment. Your pet might be relieved of a sense of duty to defend you in an unfamiliar and threatening situation and be more compliant when you go. In addition, your temporary removal might help calm your dog, as part of a pet's anxiety might be amplified by your anxious concern.

If your pet is increasingly uncontrollable at the veterinary clinic, you have several options. You might inquire about the availability of a house call by your regular veterinarian or through a mobile practice in your area. You could try going to another veterinary clinic, but your cat or dog's phobia will usually be transferred there.

Another option requires your time and energy but is a valuable investment for everyone. You can undo all the negative associations your pet has formed with regard to the veterinary clinic, but it takes time to replace them with positive associations. Retraining begins with methodically exposing your pet to the sequence of events that precede your trip to the clinic.

- If your pet only goes in the car or crate when it is time for an annual check-up, these events become reliable cues that trigger anxiety.

- Isolate these events from the clinic experience and work on making your pet comfortable with, for example, more frequent and progressively longer car rides or practice daily confinement in a cat carrier.

- Once this is accomplished, make frequent visits to the clinic only to sit in the waiting room briefly before returning home.

- Eventually, enter the examination room and ask the veterinary staff to give your pet a treat. Make sure you arrive at an arranged time when they are not busy.

This process may take many weeks and even months but everyone, including your pet, will benefit.

No one enjoys the sight of a frantic pet, least of all the veterinarian who will not only be required to handle the animal but likely will be blamed for its fears. Veterinarians usually do what they do because they love their patients and the ability to relieve their suffering. If your pet's anxiety is increasingly problematic, it would be a good idea to voice your concerns. If your veterinarian's reassurance is not enough, referral to a veterinarian that specializes in pet behavior problems may be necessary to explore additional methods. Your patience and a little time and effort will result in happier veterinary visits for all concerned.

A pet that is aggressive toward men has been beaten by one.

Behavior is a product of acquired experience and inborn predisposition. Although some unfortunate pets may be subject to physical abuse, many more pets are innately shy or fearful without prior abuse. Pets recognize individual people and other animals that become familiar to them and are capable of distinguishing males from females. Individual recognition is based in part on the perception and interpretation of odor, vision and hearing.

In general, men behave differently from women. They tend to be larger, speak in deeper voices at louder volume, and they move differently than women do. Many dogs and cats are shy of men but that does not mean they have been abused, only that they are unfamiliar with men as individuals. Exposure to a variety of different people and experiences during a critical period in a young animal's development (usually before three months of age) results in a more socially confident pet

A dog that becomes aggressive if disturbed during feeding was probably starved as a puppy.

Dogs can become aggressive about food even when they have never experienced food deprivation. Some dogs that have never skipped a meal, let alone faced starvation, can become very tense near food. Other dogs that have certainly known hunger never become aggressive at mealtime.

There are basically two types of aggression related to food. The first is guarding, or "possessive" aggression, in which a dog zealously monopolizes an object (edible or nonedible) in its possession and defends it against anyone that attempts to remove it. An animal's motivation to possess an object considered valuable is unaffected by social prominence or rank. A dominant dog, for example, will not insist on removing an object from a subordinate that is aggressively intent on keeping it. Possessive aggression reflects the value placed on the object by the dog, but can also demonstrate its capacity to become aggressive in other circumstances.

Another type of aggression related to food is the competition over natural resources. Sometimes referred to as competitive aggression, competition over food is really a special case of territorial aggression. Animals that occupy the same territory naturally compete over vital resources of food, shelter and water. Even dogs that have grown up together as littermates in total harmony can become rivals at mealtime. In extreme cases, a dog can defend a food dish and even the feeding room from housemates.

To decrease anxiety between dogs over food, separate them in different rooms if necessary, so that they need not compete at mealtime. Increasing the number of meals during the day, to two or more as necessary, can also help. Uneaten portions of food should be removed so that the more compulsive dog is not able to persist in the obsession to defend it.

Possessive aggression (guarding) occurs only with food and indicates the dog's dominance.

Dogs can aggressively defend a food bowl or other edible object such as a rawhide bone. They can also guard nonedible objects that have particular interest or value.

This behavior is not motivated by deprivation, hunger or social status. It is simply the act of an individual that seeks to monopolize an object regarded as valuable. Guarding behavior, also known as "possessive" aggression, is unrelated to the relative social rank between the guarding dog and another individual threatening the possession. The object's importance is reflected by the guard's willingness to defend it at the cost of potential self-injury. A submissive animal can and will successfully fend off a dominant animal's challenge over the object of interest. In this case, social dominance is less important than the animal's motivation to defend its possession. It may also be an insight into the aggressive tendencies of the individual in general. Some dogs are perfectly placid in most circumstances except when it comes to anyone challenging to remove their favorite toy. The fact that a seemingly placid dog can become ferocious if riled by the right trigger, can provide useful insight into temperament.

Cats can also become possessively aggressive, particularly over special food treats, but their possessive aggression is unlikely to take on the same intensity or frequency as it sometimes can in dogs. Like dogs, cats will try to withdraw from the threat to a favorite resting spot or private corner to enjoy their treasure undisturbed. In general, it is unwise to pursue an animal guarding an object of value. Your persistence will be punished. The most important thing to keep in mind is never to risk injury to yourself. Even if a pet is defending an object that is potentially harmful, you will invariably be better off to deal with this indirectly.

There are several ways to discourage guarding behavior in a dog. If your dog is a "garbage hound," secure the garbage however necessary. If your dog habitually ingests objects during walks, keep the dog on a leash and, if necessary, consider using a basket-type muzzle to avoid intestinal injury or obstruction and resulting surgery. Teach your children never to interfere when your dog is chewing or playing alone with a favorite toy or eating anything. If your dog becomes uncontrollably or uncharacteristically aggressive when given a specific treat or toy, consider withholding this object in the future. It is unwise to reinforce any type of aggressive tendency in a pet. If your dog has seized control of an object that could be harmful if swallowed or chewed, do not try to remove it from the dog's jaws. Instead, pretend you are unconcerned, at a safe distance, and try to bribe the dog away from this object by offering a substitute of equal or greater interest. This might be another food treat or the chance to go for a car ride, for example. If you convince the dog to abandon the item, remove and secure it. Of course, this type of enticement is, in a sense, a type of reward for undesirable behavior that is best used only occasionally.

Every dog, regardless of whether or not there is a tendency toward possessiveness, should probably be trained to relinquish objects on command. Begin with objects that are not of particular interest or value to the dog and graduate to progressively more problematic items. Two commands are useful for this. First, "Leave it" should keep the dog from approaching or mouthing an object that is within reach. Second, "Drop it" should cause the dog to drop any object. This is an appropriate opportunity to use food rewards with your pet for a willingness to exchange objects of value.

A pet that is aggressive toward the owner is ungrateful.

Although it is tempting to assign human values and characteristics to our pets, it is unlikely that they are capable of the same degree of complex emotions as are humans.

Feelings of revenge, guilt, ingratitude or responsibility are human concepts. The interpretation and comparison of nonhuman physical or behavioral characteristics with human qualities is called anthropomorphism.

In order to remain objective in the absence of proof positive that other species are capable of complex thinking, anthropomorphism should be avoided. Thinking in nonhuman animals is virtually undeniable, but remains difficult to prove. An animal's behavior is a reflection of individual life experience, learned and inborn. A pet's behavior is more likely based in simple cause and effect associations rather than in human notions of projected emotion and intent.

A dog that becomes aggressive toward the owner's mate is exhibiting jealousy.

A dog can react aggressively when the owner is touched affectionately by a new friend or even a spouse. The dog's behavior is unlikely a sign of any emotion like human jealousy, but is more likely triggered by canine concerns. Defense of pack members and territorial intrusion can combine to form a nonspecific reaction to unfamiliar behavior. The dog might perceive the physical posturing of affectionate advances between adults as a dominance-related challenge to the social status of a pack ally. It is unlikely that dogs recognize any elements of human sexuality. Before we assign any human frailties or faults to our canine companions, we should remember that a dog views the world with the mind of a dog. We, on the other hand, view the world with the limited capacity of our species.

A pet that suddenly becomes aggressive is having a seizure.

The sudden onset of aggressiveness in a previously unaggressive pet should be of concern to you and to your veterinarian. Although most seizures are neuromotor rather than emotional episodes, unusual behavior can be the first indication of a medical problem.

An animal that is not feeling well may demonstrate a sudden change in temperament. When the possibility of a medical problem is eliminated with reasonable certainty, ask your veterinarian to refer you to a veterinarian specializing in pet behavior problems. Aggression often seems to occur abruptly, but careful tracing of past behavior usually reveals that the problem emerged gradually. Extreme aggressive outbursts in English Springer Spaniels have been referred to as a type of epileptic "rage." These cases, seen in many species, probably represent explosive forms of dominance aggression. If the aggressive problem is in a mild and uncomplicated form, Obedience training may provide a simple solution. More severe cases should be referred to veterinary experts.

A pet that suddenly becomes aggressive probably has rabies.

The virus that causes the disease known as rabies can, indeed, cause physical symptoms which include uncontrollable attacks by the infected animal. Rabies does not always develop into the stereotypical image of a crazed and frantic creature, frothing at the mouth and lunging at anything that moves.

Symptoms of rabies generally fall into two categories. In its final stages, the "rage" form includes abundant production of saliva, aggressive behavior, restlessness and a frenzied appearance. The "paralytic" form is far more subtle and can progress to isolated muscular tremors or the paralysis of a single limb.

Both sets of rabies symptoms overlap with many other illnesses, and a diagnosis may be pronounced only on confirmation of the disease by autopsy.

Direct contact with the body fluids, principally saliva and blood, of a rabid animal is the most common way of transmitting rabies. The rabies virus is one of the few viral diseases that is not exclusive to individuals within a species or limited to related species. All mammals can be infected by contact with rabid individuals of any other species. The disease is fatal, (although people have the treatment option of a series of injections to prevent the virus's progression).

The form the disease takes is determined, in part, by the species of the infected animal. A cow that is bitten by an infected raccoon, for instance, is more likely to develop the "paralytic" form of rabies while the infected raccoon is more likely to display signs of "rage."

Another important factor that affects the disease's progression is the point of entry of the virus. An animal that is bitten in the toe, for example, may be slower to exhibit the disease than another that has been bitten in the arm. The longer the path toward the central nervous system, the slower the disease will progress and the more vague the associated physical signs of illness can be. Pets suspected of possible exposure to rabid animals may be quarantined from ten days to six months depending on their own vaccination status, municipal or state laws and the knowledge that the disease can progress very rapidly or at lethal leisure.

There are many reasons why an animal can become aggressive even in the absence of an apparently justifiable trigger. Acute physical discomfort, for example, caused by the passage of a kidney stone or intestinal cramps, can cause transitory aggressive behavior. Dominance aggression or extreme irritability in individual dogs can be unleashed suddenly and with great intensity during apparently innocent interactions with people. Epilepsy that centers in specific emotional centers of the brain can cause emotionally explosive seizures. The sudden and uncharacteristic outburst of aggression in a pet is not necessarily rabies, but warrants veterinary investigation nonetheless.

If you suspect your pet has come in contact with either a wild or domestic animal that seemed to be ill or was already dead, do not touch it. Have your own pet examined and report the incident to your local animal control officer. Even if there is no apparent injury to your pet, remember that rabies can be transmitted by direct contact through any microscopic tear in the skin. Make sure that your pet's rabies vaccination is kept current according to your veterinarian's recommendation and your state or provincial laws.

Calico cats are more aggressive than cats of other coat colors.

The calico cat has a tricolored coat with patches of black, white and orange or blue, cream and white. It is a color that is linked with the female gender and in very rare cases, a male. A male calico is infertile. Calico cats are no more likely to be aggressive than any other cat. There are no scientific data to substantiate the association of any temperament characteristic with coat color in the cat, or dog for that matter. Some studies seem to suggest, in species other than our domestic pets, that a black coat may be associated with a more tame individual but this remains unproven, too.

A cat will become aggressive or more likely to bite if declawed.

There is no evidence that a cat will suddenly become aggressive or that a previously aggressive cat will be more aggressive following surgical removal of the claws.

The declaw surgery will not cause a cat to bite rather than to scratch. Regardless of a cat's age at the time of surgical removal of the claws under anesthesia, temperament will not likely change as a result of the surgical procedure. An already aggressive cat under specific circumstances prior to any surgery will continue to be aggressive in those situations after surgery.

The surgery to declaw involves the amputation of the last joint of each digit. This is always done under general anesthesia. There is pain for the first few days following surgery, and so most veterinarians require cage rest in the hospital. By the time the cat is sent home, the cat is almost fully recovered. Some tenderness may persist for several days or even weeks in a cat that is declawed as an adult. This procedure is not advised for outdoor cats (and cat owners are not advised to allow their cats to roam freely outdoors).

If you are still undecided about whether or not to declaw an indoor cat, it is worthwhile to make every effort to promote the use of scratching posts. This should begin from the moment a pet cat is introduced to your home, regardless of age, although kittens are quicker learners. Scratching posts or boards come in many shapes and sizes. Cats are individuals with an individual preference for the texture and location of scratching surfaces. By trial and error, owners are bound to find the type of scratching post and location that suits their finicky cat.

The decision to declaw a pet cat is the owner's decision, although it is increasingly unpopular. Despite the controversy, it seems preferable to declaw a cat, despite some postsurgical discomfort, in favor of a long and happy life. Total refusal to consider the procedure could result in a resentful and frustrated owner and an unhappy, rejected or euthanized pet.

Petting or brushing a pet against the direction of coat growth will cause aggression

(See chapter 14, "Grooming," section entitled *Petting or brushing against the direction of coat growth will cause aggression.*)

Dog fights and cat fights are to the death.

Most fighting between dogs or between cats are usually brief and the injuries inflicted are most often minimal. Conflicts arising between rival males can be the most intense, and injuries can be more severe. This does not mean that death will result from a fight between males but that the opponents are less likely to inhibit the harm they inflict on the other. Fights between male cats or male dogs also can be brief if the victor is decided early and the conquered chooses not to immediately challenge the outcome.

Fights between females can result in injury as well, but in most cases females are less combative than males. Maternal aggression, the defense of the young, can be the exception. A mother cat (queen) is not as actively defensive of her young as is a mother dog (bitch). This does not imply that dogs make better mothers than cats but may reflect a significant physiological

Dogs and cats can coexist happily, especially if they are raised together from a young age.

difference between them. Because the normal queen can come into breeding condition every few weeks, she has more chances at replacing a lost litter. The normal bitch is receptive and fertile to a male dog only twice a year, and so has more invested in each litter she produces. This must certainly impact on the attitudes of these species and maternal defense.

All dogs hate cats.

Actually, some dogs love cats. A dog can tolerate the silliest antics of a kitten walking all over them. Dogs have been known to be almost nurturing of cats, grooming them as if they were puppies and playing with them as if they were another dog. Dogs not exposed to cats from a young age should be introduced to cats very gradually and under strict supervision, or not at all. Although dogs can be tolerant of cats even without prior experience, there is a risk that the dog's predatory instincts will be instantly aroused. The risk could be lethal to the cat, so make sure it is a carefully thought out one. Dogs can tolerate cats that share their household and still menace cats they encounter elsewhere. Some dog breeds are more consistently aggressive toward cats than others. The Akita and Chow may be less trustworthy with cats unless they are raised together from a young age, but a dog of any breed should not be considered inoffensive without a controlled test of inborn predisposition.

All cats hate dogs and will scratch out their eyes.

Cats defend themselves according to circumstance and experience. The cat is far more likely to avoid an animal perceived to be a threat.

If the frightened cat is cornered by a dog, whether the dog approaches the cat out of curiosity or with the intent to harm, the cat may use any weaponry for defense and try to escape.

The protruding muzzle of many dog breeds is useful to block the blows of an agitated cat, (although this is not the reason that some dogs' muzzles are long). Dogs with blunted facial structure tend to have more prominent eyes and may be at higher risk of eye injuries in general. A Pekignese that is pursuing a cat may be more likely to suffer a corneal scratch, but not necessarily because the cat was aiming for the eyes. A defensively aggressive animal will strike out at any available target although many animal species do target specific physical vulnerabilities (including the eye) for defensive or offensive purposes.

Cats do not seek out and pursue animals that are bigger than they. Cats are not malicious or evil creatures intent on blinding dogs. They are intelligent and cautious animals that usually have keen instincts for their own survival. A dog that suffers a scratch inflicted by a cat on the eye's surface (corneal laceration), will heal rapidly with veterinary attention. The cat's injuries from the same dog would be less easily reversed.

Some cats have no fear of dogs and can even develop close and enduring bonds with dogs that are similarly inclined. Other cats, even those that are raised with dogs, may never develop a real bond with their canine housemates and will tolerate them only from a distance. Cats that are unfamiliar with dogs occasionally fail to recognize danger and fail to defend themselves. Without exception, the consequences are potentially far more serious for the cat than for the dog.

A day at the beach: Dogs sometimes dig just because it's fun! (photo courtesy of Estelle Schwartz)

8

Destructiveness

Dogs that dig are looking for buried bones.

Digging is a natural instinct common to many dog breeds. Some dogs do not dig, while others seem to enjoy creating a lunar landscape on your well-manicured lawn.

Terriers, for instance, were largely bred for abilities to tunnel for prey in hidden burrows and underground nests. Dachshunds were bred for a peculiar physique to facilitate pursuit of their prey in narrow subterranean passages. Because some breeds were prized for digging does not mean that all individuals of these breeds will become problem diggers. A dog of any breed can learn to dig almost compulsively.

Dogs dig for many reasons:

- Digging in the dirt exposes a cooler layer of soil in summer heat.

- Digging in the snow prepares an indentation to help preserve body heat.

- Digging is also good exercise for the forepaws and shoulders and focuses attention on a specific goal-oriented task.

- In young animals, digging is a form of investigative play, providing the juveniles with information regarding the textures and odors of the world around them. This becomes invaluable later in life, during hunting, for example.

- Digging can enable the dog to hide partially consumed prey or bones for safekeeping.

Our pet dogs may no longer need to be expert diggers for vital survival skills, but the behavior can surface and persist as a recreational activity.

Cats also dig instinctively to uncover prey, but their reach into a burrow with a front paw is aided by their front claws, flexed and ready for hooking and retrieval. Cats typically dig before the deposit of waste and afterward to cover it. Some cats do not cover their waste or may direct their digging to adjacent surfaces. This could indicate an aversion to soiling their paws in soggy litter boxes, but most often it represents the individual's version of normal. The partial inheritance of instinctive behavior can be displayed as incompletely inherited sequences of behavior. When applied to horizontal or vertical scratching surfaces, a modified digging motion permits the cat to shed outer layers of the front claws, but this is not the same as intentional digging into the surface.

Digging occurs only outdoors.

Digging by pet cats and dogs can occur both outside and indoors.

Dogs may be particularly destructive if digging is a symptom of separation anxiety. Digging behavior may be motivated by an anxious effort to break through the barrier of a door, window or wall. Separation anxiety can include behavioral components that reflect a dog's intention to break free from confinement or to overcome an obstacle that blocks an escape route, presumably a path toward the owner.

Digging by cage-confined dogs may be an attempt to attain liberation and, when directed toward unyielding surfaces such as the floor, also reflects their frustration and redirected anxiety. Digging is occasionally observed in dogs prior to vomiting or diarrhea and so in these cases may be an indication of abdominal distension or discomfort.

Digging can take an obsessive-compulsive intensity during peak periods of daily activity. This undesirable pattern of behavior occasionally suggests that no other appropriate outlet of intellectual and physical energies is available. Many carpets bear the marks of this behavior, as does the polish of hardwood floors and the paint of door jams and walls. Stereotypic digging, an excessive compulsion to dig, can be displayed outdoors as well. Most dogs that dig to this degree would benefit from alternative activities that present them with greater intellectual challenge as well as more exercise and positive social interaction with their owners. More frequent leash walks and the daily practice of Obedience skills provide additional activity. In addition, it is frequently necessary to prevent a dog's opportunity to reach diggable areas by paving over part of the yard, or perhaps by limiting the dog's access to unpaved surfaces with a fenced-in pen.

House cats sometimes dig in potted plants. In kittens and young adults, this may be a form of investigative play. Like many young animals, including children, playing in the dirt can be entertaining. Some cats dig in house plants because they develop a preference for natural soil over the litter filler in their cat box. Should this happen, remove the target plant or place it in an inaccessible room. When this is not possible, cover the soil with aluminum foil or even chicken wire so that the cat can no longer enjoy the texture of the soil.

A pet that is destructive when left alone is acting out of anger or spite.

Actions born from complex emotions such as jealousy and revenge are part of human experience. Our pets may possess thinking minds, but they are likely not able to process the concepts that generate malice and retaliation. Spite and revenge are uniquely human distinctions, at least compared to the simpler emotions and relatively less complicated behavior of pets.

We may aspire to superiority over the rest of the animal kingdom, yet these human emotions degrade us. Indeed, we are reminded by the simplicity of our pets that we have yet to achieve the potential of our own humanity. This might give you more patience when you return home to find your puppy gnawing happily on your favorite slippers!

A pet that is destructive when the owner is away may be anxious. This might stem from social isolation and, more specifically, separation from an individual of social significance. Separation anxiety is strikingly similar across many species, including people. It is a fundamentally simple emotional reaction that can be displayed in a variety of complex ways. The void created by your absence can be overwhelming to a devoted pet, particularly if their needs for exercise, play and attention have not been filled before you left. Pets become very familiar with our daily routines and intimately attuned to the sequence of our behaviors (which tend to be constant and predictable) as we prepare to set out for work or to run an errand. Your pet's anticipation of your absence begins before you are even gone. By the time you leave, separation anxiety is at its peak, having reached this level in the first few minutes of being left alone.

Whether a destructive pet is motivated by desirable or undesirable emotions, it is very important to ensure that your pet is satisfied to rest in complete contentment by the time you go anywhere. Begin progressively longer periods of separation from you when your pet is still young. Take your dog for a nice walk to give it every opportunity to empty bowel and bladder prior to your departure. This also provides the exercise to release as much energy as possible, by directed play and even Obedience practice during the walk.

Cats, too, will benefit from one-on-one interaction in the form of play or grooming. Even so, you may find it necessary to limit access to a restricted area of your home until your pet seems better able to cope with brief periods of social isolation.

Because pets are attentive to your departure preparations, it can be helpful to rearrange the sequence of your routine. If your pet cannot identify the cues of your departure as predictably as before, anxiety will be lowered. Give your pet a special treat or toy that becomes a benefit associated only with your absence.

Avoid long and drawn-out speeches to your pet explaining the purpose and duration of your absence. They are unlikely to understand your words and may only feel more anxious because of the concern and sadness in your voice. When you return, your greeting should be calm and soft-spoken so that the contrast between absence and return is minimal. This will help your pet avoid erratic emotional peaks and valleys and, therefore, a more constant emotional level that will create a greater feeling of security regardless of where you are.

Destructiveness, particularly in young pets, can be a sign of misguided yet healthy energy and playful curiosity.

Destructive behavior is always due to separation anxiety.

Property damage caused in your absence may be attributed to several causes: separation anxiety, accidents associated with unsupervised play or even territorial defense. Not all pets are devastated by solitude, yet destruction may still occur. If a pet is left without owner participation or supervision, it is plausible that some damage may be accidental and associated with play. Young animals in particular have boundless energy which seems to come in bursts, propelling them in wild chases over furniture and madly careening through your home. Breakable objects may be sent flying during these charges, but they can also be more intentionally prodded off surfaces by curious pets during investigation of forbidden surfaces. A pet may occasionally be frightened or aggressively aroused by an intruder on your property. Isolated incidents of destruction could be evidence of a pet's attempts to defend territory, for example, by scratching or chewing through doors or windows presumably to reach the threatening individual. On the other hand, a fearful animal may destroy property in desperate attempts to escape or avoid detection.

When you are not nearby, your pet is free to explore areas that are otherwise off-limits. It is a sensible precaution to "pet proof" your home. Remove fragile objects from easily reached surfaces.

Keep in mind that cats, and some dogs, will climb onto higher surfaces. Resourceful and determined dogs will climb or jump onto chairs and table tops to reach something that attracts their interest. For cats, climbing and jumping are common talents that are part of a cat's physical as well as intellectual nature. Cats investigate elevated surfaces because they can. These agile and light-boned athletes are superbly adapted to use all the territory that is available and that includes ground-level as well as vertical space. Cats, in particular, instinctively seek the additional security of raised perches during rest. House cats can frequently be seen enjoying a catnap atop the refrigerator despite their owners objections. Destructiveness can occur accidentally during investigatory play. This can appear rather intentional when a paw is used to slowly push an object off a shelf, but this is done to investigate the object (or to attract your attention) and not out of malice.

Destructiveness is the only sign of separation anxiety.

Separation anxiety in pets generally falls into three categories: destructiveness, inappropriate elimination and excessive vocalization. Destruction of property is just one category of a group of misbehaviors that imply a pet's uneasiness when left alone. Scratching at walls, doors and floors suggests a dog's attempt to get outside or at least break out of solitary confinement. This could also be an intentional effort to pursue the owner or simply a need to urinate or defecate outside. Scratching could be part of an escape attempt. Shattered objects toppled from their original location on a table or shelf may have been overturned accidentally by a playful but clumsy pet during exploration. Objects may be broken as an animal attempts closer examination, or they may knocked down just because it is something to do.

Boredom is probably one of the main reasons that pets come to associate our absences as an unpleasant experience. Energy that is not diffused by your pet in desirable and enjoyable activities before your departure will be expressed in less desirable manners when you are away. Spend time playing with, brushing, petting or walking your pets when you are at home. By the time you are ready to leave them they should be content in every way. Your absence can become an opportunity to rest in preparation for more positive attention when you return home.

Cats use a scratch post to sharpen their claws.

Each nail on a cat's paw grows in length toward a sharp tip as well as in layers of thickness. The top layers become brittle and dry with exposure and wear. When a cat scratches against a surface, the outermost dulled layers are removed to expose newer layers beneath. The nail is not quite sharpened, rather it is essentially peeled.

Cats scratch in order to keep their claws ready for the specialized functions they were intended to serve. A cat's claws are used in self-defense, hunting, feeding, territorial marking and grooming. Both efficient tools and weapons, a cat's claws are most efficient when kept well groomed by daily use of a desirable scratching surface. Cats that are permitted to roam outdoors should not be declawed, since it will affect their ability to defend themselves against attack and may restrict their ability to avoid danger. Some declawed cats, however, adjust so well that they are still able to climb trees. This should not be put to the test for the average cat in a life-threatening situation. A cat's claws may be trimmed every few weeks. This will blunt the sharp tips to minimize some of their effect but will not deter the cat's natural inclination to scratch, regardless of the reason.

Cats scratch where they shouldn't because of owner-directed anger.

Cats scratch in order to remove dead surface layers from their claws. They also scratch to leave a visual marker of their presence. Scent glands in the foot pads advertise their scent to others to assert their territorial claim. Cats

A cat's scratching serves many functions and, although it can become destructive, it has nothing to do with spite, anger or revenge.

scratch in order to stretch their front legs and upper body. Stretching is important to tone muscles for both play and predation. It keeps muscle fibers aligned and ready for the swift movement for which cats are so famous.

Cats scratch because it probably feels good and because it is their instinct to do so. Cats do not scratch to please us or to seek revenge for anything we did or did not do. Indirectly, however, if a cat is anxious about something, scratching may help to alleviate some anxiety. If cats feel anxious, security might be increased by reaffirming an identification with their territory. Because scratching is one way a cat marks territory, this behavior may increase in times of stress. Cats tend to scratch surfaces near favorite resting places or food dishes. If you notice that your cat has chosen an undesirable surface or location, use this information to your advantage and immediately place a scratch post there.

Kittens or newly acquired adult cats should be encouraged to scratch on scratch posts or boards from the moment they are introduced into your home. Understanding that the incline, texture and location of a scratch post is an individual preference will help direct you toward improving your pet's use of a scratch post. It can be very helpful to provide a selection of different scratch posts at a variety of locations to determine which one your cat prefers. Alternate scratch posts of different textures (cardboard, sessile, carpet) at different locations and placed at different angles until your cat demonstrates which combination it finds most attractive.

Cats instinctively use scratch posts and do not have to be taught to use them.

Cats instinctively scratch tree trunks and other rough surfaces found in nature. Scratch posts are man-made alternatives intended to be the next best thing. Cats will be inclined to scratch surfaces that best offer the elements of their individual preferences.

Cats scratch horizontal and vertical surfaces as well as inclined objects at different angles. Location and angle of scratching surface may be more important than the texture of the surface itself. The corners of sofas, for example, are common targets by house cats for this reason. Other indoor targets fulfill all requirements, including a preference for nubby or irregular textures found in rattan chairs and carpets. It can be helpful to play with your kitten or cat near a scratch post and playfully scratch the surface with your fingers. Your cat will often respond by scratching the post too, initially attracted by the movement of your fingers and the sound of scratching. Some owners have found that attaching a toy to the top of post will entice their pet to reach up toward the object and thereby encourage use of the scratching surface. When you see your cat scratching at a scratch post, be sure to give abundant praise. Let the cat know that scratching at the post is the best thing your cat could possibly be doing to please you. If your cat seems uninterested in the post at one location, move it to another. Move it directly in front of that sofa corner or rattan chair. Understand their individual preference for scratching location, angle and surface texture. Watch what your cat is trying to show you, and then offer your cat the opportunity to earn your praise!

Scratch posts must be treated with catnip to be effective.

The application of catnip scent to the surface of scratch posts can indeed be useful in attracting some cats, but it does not work for all cats. Kittens younger than six months of age or so may not be able to detect catnip, and catnip-treated surfaces will be irrelevant to them. After puberty, all cats are sensitive to the chemical effects of catnip's primary active compound, nepatalactone. The initial approach to investigate catnip is under voluntary control. In other words, the cat may choose to avoid contact or perhaps to inhibit the response. This might happen because of nearby distractions, for instance, or anxiety due to territorial conflicts. Some cats actually seem to be repelled by the scent and avoid catnip even when it is available to them.

Training a cat to use a scratch post can certainly be accomplished without the use of catnip. Far more important considerations are the age at

Scratch post training is made easy by placing the post at a location that attracts this form of marking behavior.

which the cat is introduced to the use of scratching boards and the cat's individual preference for surface texture, angle of placement and location. For adult cats that enjoy the catnip effect, the scent of catnip may indeed be useful in initiating the investigation of a new scratch post. Catnip may also help because part of the behavioral sequence typical of the catnip response is scratching or digging at the catnip source. Relying on catnip alone to train desirable scratching habits in pet cats, however, frequently results in disappointed and frustrated owners. Many cats will respond just as well to an owner demonstration of scratching with their own fingers on the post for encouragement.

It is always cruel to declaw an indoor cat.

The declaw surgery (onyxectomy or onychectomy) is always performed under general anesthesia by qualified veterinarians or by certified surgical technicians under direct strict veterinary supervision. Typically, only the front claws are removed, although in extraordinary cases all four paws may be declawed. Cats who have undergone this surgery typically remain at the veterinary clinic for up to several days where they can be confined to strict cage rest. This promotes more rapid healing. By the time they are discharged to their owner's care, most declawed cats have recovered well. Cats declawed as adults may require one or two weeks more to adjust or until any remaining tenderness disappears.

There is no question that the declaw procedure is painful in the immediate postoperative period. Most surgeries are painful. This surgery, however, has become the subject of heated debate. Many veterinarians and animal rights activists consider declawing to be an unspeakable act of cruelty. Others protest that it unnecessarily disfigures the cat since it slightly changes the arch of a cat's toes. Others consider the procedure to be inhumane and protest that it must be avoided under any circumstance. Some warn of dire and tragic consequences if a cat is declawed.

The only sad consequence of declaw might occur if a declawed cat is still allowed to roam outdoors. A cat that is declawed should be strictly confined as a house pet. The disadvantage in self-defense and climbing to escape danger seems obvious.

A policy of moderation has its merits in most cases of controversy, regardless of the subject. A cat that scratches on undesirable surfaces can destroy an owner's property and, in the process, the owner's affection. It is difficult, perhaps impossible, to compare the value of property with the value of a cherished pet. But that is not the point. Just as a pet has the right to a happy life, so do owners. We each work hard for our income and a comfortable and attractive home. To return at the end of a stressful workday to find a new sofa shredded or pretty wallpaper lacerated does little to endear a pet or make us feel welcome in our own home. A pet that has become a detriment cannot be enjoyed. A destructive cat will be quick to sense rejection and any sense of well-being will in turn deteriorate.

The decision that guides whether or not to declaw an indoor cat must be based on the long-term benefit of both pet and owner. If attempts fail to train a cat to scratch on appropriate scratching surfaces, declawing is a viable alternative for a frustrated owner. Consider which is more cruel, the increasing resentment of and alienation from your pet, or temporary postsurgical discomfort in view of living happily together for a lifetime.

Trimming a cat's nails will increase scratching at undesirable locations.

Cats do not scratch in order to "sharpen" blunted nail tips. There is no evidence to substantiate this claim. The action of scratching helps to shed each worn layer of toenail to reveal an already sharper one beneath. Trimming nails will blunt the tips. The cat will scratch as much as before except the surface will not shred as effectively. At least not for several weeks until another sharp nail is revealed as outer layers are removed.

Cats scratch for a variety of reasons. One of these is to leave visible marks to stake territorial claim. The action of scratching also leaves odor markers by virtue of scent glands in the foot pads. Another important function is to stretch muscles in the forelegs, shoulders and back to keep good muscle tone and to stay ready for quick and agile motion. In case of danger, fast escapes can be important for survival. In case of hunting opportunities, agility and speed can bring obvious benefit. Periodic trimming of a cat's front nails can be helpful damage control for a pet that scratches people even with playful intention. A cat that uses undesirable surfaces may cause less destruction if claws are periodically blunted.

Cats that suck or chew on cloth surfaces were weaned too early.

Excessive mouthing of inappropriate objects is a form of oral stereotypic behavior similar in nature to obsessive-compulsive patterns in people. It is not a consequence of premature weaning, nor is it more frequently associated with orphaned kittens that are hand raised.

Oral stereotypes are seen predominantly in Siamese cats and some Abyssinians. Individuals of mixed Siamese parentage and Siamese-derived breeds also show a higher incidence. A genetic predisposition is likely, although not every Abyssinian, Siamese or Siamese-related cat will become orally fixated in this way. Breeders of Siamese may purposely extend the time litters of kittens spend with their mothers before placing them in homes. There is no evidence that this prevents the pattern, and predisposed kittens eventually show this behavior regardless of their age at weaning. Some cats begin to suck on cloth items during early adolescence. In addition to sucking, the cat may also chew the surface with the molars and even swallow pieces of fabric. While wool is a favorite material, items made of cotton and other

Wool sucking, more common in Siamese cats, may have a genetic basis.

cloth can attract the cat. This oral vice tends to generalize to a variety of targets, including items of clothing, rugs, coats and upholstery. Fortunately, the behavior tends to decrease as the cat matures.

This undesirable behavior is best nipped in the bud. As soon as a young cat demonstrates prolonged sucking or chewing on a given item, immediate intervention is advised. Intentionally baiting the surface with a substance, for example, will cause an unpleasant experience directly associated with sucking. In this way, the object itself (and not you, the owner) becomes the source of punishment. Hot pepper sauce has been recommended.

In some cases, the object may be sacrificed and the cat encouraged to direct attention toward it. The reasoning is that if the object is removed the cat might just choose an even less desirable alternative. Completely removing a favorite object may backfire and unintentionally promote the cat to graduate to a multitude of even less desirable substitutes.

Increasing the cat's activity in general, feeding several small meals a day, and increasing dietary fiber have been advised. If the behavior becomes very destructive, daily treats of the small-boned tips of chicken wings may become an alternative. The risk of intestinal obstruction from small pieces of bone are slight compared to the greater risk of habitually swallowing bulky fabric. Sedatives and tranquilizers are usually of little benefit, although a veterinarian specializing in pet behavior problems may be able to suggest complements to the benefit of additional distractions.

9

Elimination Problems

Paper training is the best method to house train a puppy.

Training a pup to urinate and defecate on layers of newspaper spread over the floor is no longer recommended. Even if the pup successfully learns to void only on newspapers, the goal of most dog owners is to teach voiding outside. One of the greatest risks to this method is that the pup will learn to eliminate indoors. Paper training is an unnecessary step that makes the training more confusing and lengthy for you and your puppy. Instead of training the dog to first use newspapers and then use the great outdoors, why not skip step one altogether? To promote house training:

- Feed your puppy (and adult dog) distinct meals at regular intervals.

- Remove any uneaten portion after your pet seems full and moves away from the food dish.

- Take your dog outside for a walk on a leash within fifteen minutes of each meal when the need to defecate and urinate will be predictably high.

Swallowing initiates peristalsis which are waves of rhythmic contractions that begin at the back of the throat and travel along the entire length of the digestive tract. Yesterday's meal will be digested and ready to be eliminated when a meal is ingested today. Every meal triggers the intestinal wave to move successive meals further along the digestive tract in a phenomenon called the gastrocolic reflex. Feeding a pet regular meals allows the owner to synchronize walks with predictable sphincter functions. A pet that has continuous access to food will have frequent and unpredictable habits and house training will be frustrating if not impossible.

Keep your puppy or adult dog on a leash when you go out for a walk even if it is within the confines of your yard. You must be close by to herald the arrival of urine or stool with immediate jubilation. Properly timed positive reinforcement is essential to teach a pet desirable behavior. Walking your dog off-leash is illegal and a safety risk, and may create other behavioral problems which include roaming (running away), eating rocks and other unhealthy objects (pica) and coprophagia (the ingestion of feces).

The best way to punish house soiling is to stick your pet's nose in its own waste.

Verbal or corporal punishment is never a constructive way to discourage inappropriate elimination.

If you discover an "accident" even seconds after the act, it is too late for your pet to connect the action to your (delayed) reaction. Sticking your pet's nose in the waste is unkind and ineffective. Mistreating a pet while you are angry will only teach fear of you and could even produce an additional "accident" out of fear.

The best method is to emphasize the positive and ignore the negative. If you catch your dog in the act, gently but firmly show that you are displeased, but forget your frustration as soon as it is expressed. Then with great enthu-

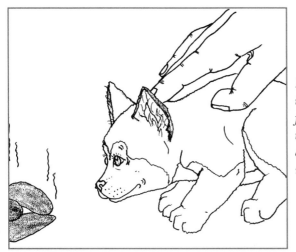

Forcing a pet's face into its own waste is an ineffective method of house training and an example of unnecessary punishment.

siasm, bring your dog outside immediately for a leash walk. If you discover the "accident" only, save your energy and simply clean it up. Disinfect and deodorize thoroughly to discourage your pet from returning to this spot again. Develop the habit of walking your dog on a leash within fifteen minutes after each meal and for as long as it takes to produce urine or stool. When either of these occurs, let your pet know that this must truly the most wonderful thing that you have ever seen!

Inappropriate elimination in an already housebroken pet may point to an underlying illness or emotional upset. This is obvious in cases of diarrhea or bloody urine, for example, but sometimes the only symptom of a latent medical problem is a subtle change in toilet habits or other behavior. If you suspect your pet is ill, contact your veterinarian with the details. If a physical cause is not determined, a referral to a veterinarian specializing in behavior problems can prevent new problems from becoming old ones that may be slower to respond to treatment.

You can tell there has been an "accident" because your dog "looks guilty."

Guilt is a human experience that results from long and careful training in morality and duty (religious or secular). Dogs do not have complicated human reactions such as guilt. It is also unlikely that they can make a clear association between house soiling and your reaction after any delay. On the other hand, our pets are expert at reading our facial expressions and body postures. By studying our behavior, they can learn to anticipate punishment or praise. If you think carefully you might realize that your expression could have conveyed to your dog your anticipation of finding an "accident" upon your return home.

The next time your dog greets your return home in a submissive body posture (head held low, ears down, body crouching close to the ground and tail between the legs) and a fearful facial expression (avoiding eye contact by sideways glances, tensed brow), think about how you appear. Your dog is not feeling guilty, but has learned to anticipate punishment from times when you did come home to an "accident" and immediately issued punishment. Instead of upsetting yourself and your pet for something the dog probably doesn't remember anyway, try to understand what might have happened while you were away and work on preventing the problem. One of the nicest things after a hard day out in the world is to come home to a dog's happy welcome. Don't deprive yourself or your dog of this pleasure.

When urinating, female dogs always squat and male dogs always stand and lift a leg.

It is true that most female dogs squat to urinate and most male dogs urinate while standing. On the other hand, a female that usually squats will occasionally stand and a male that stands to urinate will occasionally squat! Pups of either sex usually squat until puberty begins around six months of age. The position that a dog adopts to urinate is partly under hormonal control, however, it is also influenced by other factors.

Dogs urinate not only to relieve themselves but to mark territory. After thorough investigation of other surface odors, urine marking is typically deposited in small amounts distributed at many locations. Because dominant dogs, male or female, tend to be more determined in their territoriality, territorial urine marking is more effectively performed while standing. The choice of surface that is intended to be marked will also determine the dog's technique. Some dogs have even been seen to stand on their front legs in a "handstand" to reach an elevated target. Physical discomfort due to injury or arthritis, especially in the hind limbs or lower back, will also influence the pet's posture.

Male dogs need to urinate more frequently than females.

A dog's bladder size is proportional to the body and is not determined by gender.

In other words, males do not have smaller bladders compared to females of the same body size. Males do not need to urinate more than females, but the frequency of urination is determined in part by gender-related function. In general, males tend to patrol larger territories compared to females and so tend to mark slightly more. Urine advertises a dog's presence to other dogs in the area. This permits them to associate a "face" with a recognized body odor when they meet.

Females mark more frequently when they are "in heat." Urine, containing sexual signals called pheromones, tantalizes males and directs them to the attractive female. After investigating a scent marker, dogs of both sexes urinate over it. This covers a rival's odor with their own and also allows them to replace faded markers with fresh "calling cards."

In many species, including dogs, the function of waste deposit is not solely related to emptying the bowel or bladder. Urine and stool are important markers which identify the individual's claim to territorial boundaries. Both males and females mark their territory with voided waste and with other scent

Dogs investigate each other's identifying scents in territorial urine markings. Never eat yellow snow! (photo courtesy of Estelle Schwartz)

glands on various body surfaces. Urine marking, voided as small quantities of urine, differs slightly from urine emptied from a fully distended bladder although every scented drop has secondary territorial value. In both male and female dogs, just a few drops of urine are deposited in standing or crouching positions to best access flat or inclined surfaces.

In cats, both male and female can squat or spray when marking with urine, leaving minute quantities of aerosol droplets or more generous samples. This can give the illusion of a frequent need to urinate in either sex but should not be confused, for instance, with an underlying bladder infection. The discomfort of a bladder infection can cause symptoms of an increased urge to urinate, producing only a drop or more copious quantities. In urine marking, there is no discomfort and the animal controls intentional placement of scent.

A dog that leaks urine in greeting has defective sphincters.

Meeting someone for the first time can be stressful. To a young pup, or an adult dog who is socially insecure, greeting another dog or person can be intimidating. Even when the visitor is familiar and friendly, the anxious dog, and urinary sphincters, may be overwhelmed! Urine that is voided in the context of social interaction suggests that the dog is unsure, perhaps even fearful, of how it will be accepted.

The submissive dog urinates to greet an unfamiliar individual or some-one already recognized as being socially superior or dominant. Submissive urination also signals that the dog intends no threat. Of course, it takes time for many young pups to develop voluntary control over their urinary sphincters. Individuals mature physically and psychologically at their own pace. By the age of six months, however, most pups are able to "hold it." Submissive urination is quite common in timid puppies and is more frequent in females than males.

Individuals take time to develop self-confidence and trust in others. Understanding that submissive urination reflects a dog's social uneasiness should convince owners to be patient and avoid gestures that intensify the dog's anxiety. Greetings should be calm so as not to overwhelm a shy dog. Keep your voice low and tone of voice gentle. Avoid staring directly at or leaning over your pup since these can be menacing. Sit down so that your physical presence is less overwhelming. Pet your dog under the chin rather than over the top of its head or back since this is a subtle message of domi-nance which need not be emphasized to a submissive pet. Most young dogs eventually outgrow this tendency by acquiring confidence through interacting with their owners and other friendly pets.

Unlike cats, dogs do not mark their territory by voiding indoors.

The dog's territory is centered in your home and extends outside if your dog is taken for walks. The house cat's territory is defined by the walls of your home. For cats that roam outdoors, and many dogs, territorial boundaries are flexible. Both cats and dogs use urine and stool to mark territory.

House training implies that you teach a pet to void at sites only outside your home which, when you think about it, is contrary to a territorial nature. Cats at least have the option of using litter boxes indoors. For most dogs, how-ever, urinating or defecating anywhere inside the home is taboo.

Imposed toilet habits can be more of a challenge for the strongly territo-rial dog, and in particular, males. Intact or neutered male dogs can stubbornly lift their legs against the side of your sofa, for example, and seem blithely unconcerned by your negative response. Neutering can be helpful in con-trolling the hormonal urge to mark, but this behavior can continue even in neutered animals, maintained by the sight of the sofa or traces of odor. Clean-ing, deodorizing and preventing unsupervised access to favorite targets is essential. Frequent leash walks give your pet the chance to void outside and give you the opportunity to reinforce elimination in more appropriate locations.

Dogs eat stool because they have a nutritional deficit or abnormal temperament.

Distasteful as it may be from a human perspective, it is normal for a mother dog to ingest her offspring's waste. Coprophagia, the ingestion of feces, has obvious benefits for hygiene of the young and decreases the attraction of predators by keeping the den area clean. This instinctive behavior is common to many species and has real survival benefit. Mother cats are fastidious in caring for their kittens, yet only queens nursing infants practice this behavior. Coprophagia in cats does not persist beyond maternal care. This may be due to the feline instinct to cover feces and because cats are generally more selective about what they eat. This behavior not only creates pungent halitosis, but also predisposes the animal to parasite infestation.

Puppies normally begin to crawl near two weeks of age and walk by the age of three weeks. Young pups may complement their mother's efforts to keep things tidy when they begin to walk but often cease to ingest their own waste products as they are able to roam farther away. Some pups persist in coprophagia as long as it is available. This can become a persistent problem if owners are in the habit of walking their dog off-leash or do not clean up waste deposits in their backyard. Dogs might even develop a taste for stool, seeking out their own, those of other dogs and even raiding cat litter boxes and diaper bins. The consistency and flavor of feces may be part of the attraction. The quality of a dog's food directly affects digestion and, therefore, the quality of feces. A coprophagic dog may prefer soft stools while another relishes "poopsicles," the frozen kind that are a winter treat! Many methods claim to control coprophagia. Advice to change the diet, or to supplement food with products that claim to alter the flavor and consistency, frequently fail. The best way to stop coprophagia is to prevent the opportunity. The cat box should be accessible to your cat, but not to your dog. Walk your dog on a leash, remove feces immediately and move briskly away from the area. Patrol your yard daily to promptly dispose of any solid waste. After a long period of good behavior, your dog will be less likely to resume old habits, but if given the chance, may take it.

Cats do not eat their own feces because they are smarter than dogs.

Coprophagia is not reported in cats perhaps because they are more selective about what they eat. From a young age, it is also their nature to bury their waste. When newborn puppies begin to mature they may mimic their mother's instinct to keep the den clean so as to prevent detection by predators. This may complement her maternal care, but it may also be part of a pup's normal investigation of the immediate environment. As they develop, most pups will lose interest in orally investigating or ingesting stool. Some however, will persist in coprophagic behavior as long as fecal matter is available.

Coprophagia is corrected by treating stools with a chemical or distasteful substance or by changing the dog's diet.

Many methods are recommended to cure pets of this unattractive behavior. These include modifying the taste or texture of bowel movements by sprinkling them directly with special substances. A change in diet may alter the consistency or odor of stool, but this, too, is an unreliable method. The only reliable way is to ensure that all feces are completely unavailable.

Dogs should be walked at regular intervals on a daily basis. Walks should always be on a short leash so that the owner can quickly move the dog away from its own feces. Accumulated yard waste should be removed since older deposits can also become attractive. Dogs should not be allowed to explore the yard unless they are supervised and on a leash. A determined coprophagic dog can be quick and every effort must be made to erradicate the bad habit. In some cases, short-term use of a basket-type muzzle during leash walks will

Daily leash walks are needed for every dog, regardless of size, to maintain house training and to prevent problems such as coprophagia.

provide more time to move away or to collect the fecal deposit. The ingestion of feces may not be restricted to the individual's own stool. Some dogs will ingest the stool of other animals indiscriminately. Others develop a taste for the waste of other dogs or cats. Dogs that are leash walked will be more easily controlled. Dogs that raid the litter box should be prevented from gaining access to it. In some cases, it may be necessary to get creative. Place the box in a room that is inaccessible to the dog because of a restricted doorway or cat door. Elevate the box or cover it with a special hooded cover. Angle a covered box toward the wall so that the cat has just enough room to use it but not your dog. Litter boxes should be kept meticulously clean to limit odors that entice dogs.

Coprophagia is not an indication of a medical or psychological abnormality. It is generally not harmful although it can be an important source of parasite contamination. More than anything, it can cause great distress in owners who may find this pattern so repulsive that the pet is rejected altogether. Coprophagia is an instinctive behavior that is reinforced by repetition. By making feces inaccessible, the dog can unlearn the habit but may regress if the opportunity presents itself. Keep your dog well-exercised with walks on a leash. Spend time playing entertaining games that teach or challenge Obedience skills. Provide a variety of chewable toys to satisfy the need for oral stimulation in more acceptable ways.

Starving your dog will teach bowel control.

Swallowing food triggers waves of muscular contraction that travel along the digestive tract and results in the elimination of feces. If meals are eaten at irregular intervals or if the bowel is sluggish, there may be no bowel movement to be voided.

On the other hand, if a bowel movement is prepared and in position, removing a dog's food is unlikely to prevent defecation. In addition to feeding, one of the things that trigger the elimination of feces is fear or anxiety. Illness can affect the consistency of stool as well as a dog's ability to control the sphincters. Young dogs need time to be able to consistently control their sphincters and yet they also require frequent feedings. It is not advisable to restrict food in growing pups. Feeding them at reliable intervals is best.

House training is made easy when regular meals are fed. This way, natural rhythms of the digestive process can be synchronized with outdoor exercise. Once the animal is full, remove the uneaten portion of food and take your pet directly outside for a walk. By limiting access to food except at distinct mealtimes, your pet will have long portions of empty intestines corresponding to the intervals between meals. This will allow you to anticipate the dog's toilet habits so that you can both relax between meals. Your pet may still need to urinate between meals (upon waking from a nap, for example) so most dogs should have walks between meals, too.

"Scooting" is usually a sign of internal parasites.

"Scooting" is a peculiar position seen in pets who have an itch to scratch and no other way to reach it. In a modified sitting posture, the pet walks forward on the front legs, dragging the rump and anus along the ground. This behavior can be preceded or followed by an attempt to lick or bite at an adjacent body surface (such as the tail, back or feet). Scooting is a nonspecific sign of discomfort at or near the anus. This behavior is seen from time to time in normal pets that feel mildly itchy. Should scooting become excessive and associated with other signs of persisting discomfort it could indicate a medical problem.

Scooting can be associated with discomfort such as diarrhea caused by intestinal parasites, but it could also be due to many other things. Gastrointestinal upset due to viral or bacterial infection, diet change or food intolerance can cause discomfort in the area. Fleas or other insect bites on the hindquarters can also cause this behavior. The anal sacs, a pair of scent-producing glands near the anus, can become plugged or infected, causing the animal mild to severe discomfort. If your pet scoots repeatedly in a day or over several consecutive days, it would be prudent to consult a veterinarian. Annual fecal analysis is recommended even in the absence of scooting to determine the presence of internal parasites.

A pet that voids indoors when you are home is acting out of spite.

The only animal that is known to behave with malice is the human animal. When a pet urinates or defecates in an undesirable place it is called inappropriate elimination. If this occurs while an owner is away from home, it may be a symptom of separation anxiety (although other causes should be investigated). A pet that voids indoors when its owner is at home may be ill and unable to control its bodily functions.

A cat or dog suffering from a urinary tract infection may have an increased urgency to urinate and could be unable to control itself. Prostate infection in males, for instance, can affect urinary control and sometimes causes diarrhea. Frequent urination typifies diabetes mellitus and other endocrine disorders. Diarrhea can be associated with dozens of gastrointestinal diseases and many other illnesses unrelated to the digestive system. Fecal and urinary incontinence can be a sign of neuromuscular disorders.

Inappropriate elimination that occurs when the owner is home can be accidental. A pet could have made an attempt to communicate the need to go outside to a preoccupied owner. Perhaps the dog never learned to indicate this need. A recently acquired dog may not know how to communicate this need. Puppies and adult dogs can lose house training because the technique used was confusing (e.g., paper training) or not consistently reinforced (e.g., let out in yard alone vs. regular leash walks).

Inappropriate elimination is frequently related to emotional upset. For example, a pet may have been frightened by something or someone. The household could be adjusting to the arrival of a new pet or baby. An animal may be anxious because the owner is preoccupied with job-related stress or a new "significant other." Inappropriate elimination in these contexts is not jealousy, but rather an emotional response to a perceived loss of attention and a reflection of the owner's stress.

Inappropriate elimination can also be related to territorial affirmation in both cats and dogs. Territorial anxiety, and associated house soiling, can occur as a consequence to any emotional upset or psychological upheaval.

A pet that voids indoors when left alone is acting out of spite.

Jealousy, spite and revenge are exclusively human emotions not shared with other animals. A pet that urinates or defecates in the owner's absence may, however, be acting out of anxiety stemming from a number of sources.

A pet may void in an undesirable location out of fear caused by something inside or outside the home. Young pets need time to gain control over sphincters and their development varies. Incomplete house training or regression in training can occur if a pet is not given ongoing opportunities for good habits to be reinforced. Physical ailment can cause a pet at any age to lose control. Pets may simply be unable, because of the absence of care givers, to reach a desirable location. Separation anxiety, the emotional reaction to social isolation from a pet's owner, is one of the most common causes of inappropriate elimination. It can occur in a pet of any age even after a lifetime of successful house training and is a reflection of the individual's emotional or physical state at that moment. In essence, to suggest that this is the action of a vindictive and malicious pet is to misunderstand the behavior.

For a cat, litter pans must be changed on a daily basis or well before the cat's tolerance for minimal hygiene is reached. Litter pan placement and the type of litter filler must be considered.

Punishment is never productive, particularly when it occurs following an "accident." A pet that looks "guilty" when the owner returns, does not remember what occurred even moments earlier, but rather is anticipating the owner's punishment from subtle cues in facial expression or body language. Save yourself the aggravation and clean up the mess and concentrate on the steps that will make future "accidents" less likely.

Restrict or remove water to stop inappropriate urination.

Removing a dog's or cat's water bowl in order to control inappropriate urination (voided in an undesirable location) likely has little or no effect in teaching desirable toilet habits.

Control over urinary sphincters until a designated area is reached takes practice. It is necessary to be tolerant of "accidents" and to provide more than adequate opportunity to void at times when the bladder is filled. Urine is continually produced by the body, but is most plentiful shortly after a meal or a period of rest. Urine production is ongoing and removing water does nothing to stop it. Restricted drinking may reduce urine output, but urine will still be produced.

The only thing that is guaranteed by removing a pet's water dish overnight, for instance, is that the water will not be dumped out by a mischievous pet. The amount of water that a dog or cat might actually consume overnight is minimal. Dogs and cats tend to drink most after a meal or after exercise. Most dogs are far less active at night than during daylight hours and would have had little more than a few sips of water anyway.

Even if particularly active at night, limiting water intake will not convince a cat to use a litter box if it is otherwise inclined. Still, it seems unkind to deprive a pet of even a few sips of water if it feels thirsty. With very few exceptions, it is rarely recommended to restrict your pet's source of fresh water. If an animal is denied access to water in extreme conditions of heat or humidity, for example, the body will naturally attempt to control water loss by requiring the kidneys to conserve fluid lost in urine. Water intake is essential even in cold weather, and an animal can become dehydrated if fluid intake is restricted even in cold temperatures. To deny an animal water in most circumstances, particularly in extreme temperatures, is unnecessary and could result in illness. Water deprivation can be a life-threatening stress in animals suffering from undiagnosed disease or from recognized illnesses. It does not make sense intentionally to dehydrate even a healthy pet to curb the desire to urinate. (House training tips in dogs are discussed in this chapter, and tips for litter box training in cats are also described in this chapter.)

If your pet drinks what seems to be an unusually large amount of water overnight, you should consult your veterinarian. Many serious medical conditions, such as kidney disease and diabetes, include symptoms of excessive thirst. Once your veterinarian has eliminated the possibility of underlying physical problems, it might be worthwhile to consider changing to a diet that is lower in salt or sugar content (many canned foods are higher in sugar compared to dry foods), which might make your pet thirstier than usual. Psychogenic drinking, excessive thirst due to emotional stress, can occur in the resident pet with the addition of a new pet to your household. It will certainly be helpful to make sure your dog has a long leash walk every evening so that it has the opportunity to empty both bowel and bladder. It is important that you accompany your dog, as opposed to assuming it has voided in your yard, so that you are there to see and praise these actions. The emptier your pet's bladder is before bedtime, the better able it will be to control urinary sphincters overnight.

Urine marking is prevented by neutering a cat before six months of age.

Urine marking can occur in both male and female cats who are sexually intact or neutered. While sexual hormones decrease immediately after surgical removal of a male's testicles or a female's ovaries, urine marking that began prior to the surgery may continue.

Moreover, urine marking can begin even years after neutering in a pet that never displayed the behavior. Obviously, sexual hormones cannot be solely responsible for driving a pet to mark with urine. Even when the hormonal influence has been eliminated, a cat with this strong inborn tendency toward urine marking can fulfill it.

Urine marking by cats is an effective method to leave identifiable scents. Chemicals in the urine are exclusive to each individual cat and serve as potent "calling cards" to other cats. Urine markers identify individual cats with specific territories. Territorial urine marking occurs in solitary indoor cats, and this implies that urine marking has significance other than social signaling. A cat's sense of emotional security is very much linked with its territory. The very act of depositing urine on a target reinforces a cat's territorial security and, therefore, brings immediate emotional release. This explains why a cat may be more inclined to mark territory during times of stress. Regardless of why the pattern was initiated, it will be maintained by lingering odors and the self-reinforcing nature of the behavior.

It is important to determine the possibility of an underlying physical ailment in a cat that fails to urinate in the litter box. Infections or inflammation of the urinary bladder or kidneys can be very uncomfortable and cause urinating in unusual locations. Diseases such as diabetes can result in elevated urine production which can overwhelm even the most fastidious cat. Medical conditions that do not directly affect the urinary tract can cause inappropriate urination because a sick cat can express anxiety by urine marking.

A litter box needs changing only once or twice each month.

There are basically two factors that determine how frequently to change a cat's litter filler. One of these is the owner's tolerance to the litter box odor and basic hygiene. The other factor is the cat's individual tolerance to litter hygiene and odor. Of these two factors, the only one that really counts is the cat's own preference.

There are many brands of litter available. Most of these are marketed to attract the owner's attention and do not guarantee that the pet will find the product as attractive, even if used according to the manufacturer's recommendations. Although a litter label might appeal to a cat owner because instructions indicate that the litter box be changed once a week or less, it does not follow that a cat will find this appealing.

As a general guideline, use common sense when deciding how often to replace fresh filler in a cat's box. Provide one box for every cat you own. You may prefer to remove feces only and delay changing the litter filler or you may prefer to completely replace litter filler each time. Your cat's preference must always come first. Even if you have only one cat, remove fecal material at least two to three times every week. If you have two or more cats, it may be necessary to remove stools on a daily basis. If you use "clumping" litter, be sure to remove urine and stool every day. To be safe, dump the remainder of the sand every week and replace with a fresh layer in a clean box.

Regardless of the type of litter filler you select, place only enough litter filler to cover the bottom. It need not be inches deep to provide your cat an adequate quantity to cover the waste. To determine how often to completely replace litter, use your nose to guide you and remember that your cat can detect smells that you cannot.

A cat that does not cover litter box waste is abnormal.

A cat that fails to cover waste is neither lazy, abnormal nor intellectually impaired. Not all cats are born with equal instinct to bury their waste. Some cats dig furiously before and after anything they deposit. Others seem to dig more enthusiastically after producing a bowel movement while others barely dig in the box at all. Some cats displace the digging to the side of the box or the floor just outside the box. In some individuals, this modification of the digging sequence may reflect an aversion to a soiled litter box.

Cats may continue to use a box, despite the soil, but dig less or on an adjacent (and cleaner) surface. Another reason for variation in digging technique may be that the cat has a wound because of recent injury or surgery on a paw or leg. Preference for litter texture and the addition of deodorizers can also be reflected in a cat's digging behavior.

Only male cats suffer from Feline Urological Syndrome (FUS, cystitis, urinary crystals)

(See chapter 13, "Nutrition," section entitled*Excessive water intake should not be permitted.*)

Minerals in the urinary tract can form crystallized sediment in both male and female cats.

Individual cats prone to form urinary crystals are more likely to do so when fed a diet high in "ash." Ash content refers to the relative quantity of magnesium compared to other minerals in food. This is typical of seafood-based foods and most commercially prepared dry foods. In recent years, the pet food industry has begun to formulate pet foods that comply

To prevent house soiling, the ideal is one litter box for every cat to help maintain litter box hygiene and consistent use.

with veterinary guidelines for prevention of nutritional contribution to disease. Until pet nutrition is better standardized or monitored, prescription pet foods are available through veterinary clinics that prevent or treat pets with specific diseases.

Crystals are not the only cause of FUS (or feline urinary tract disease, FUTD). Viral and bacterial infections can also cause inflamed tissues. There are likely many causes of this syndrome, some of which we have yet to discover, and it is possible that more than one cause can exist in the same cat.

While both male and female cats can become very uncomfortable because of the inflammation caused by irritating bladder crystals, these rarely cause severe health problems in females. Urinary crystals are washed to the outside in streams of urine. In males, the passage of the urinary tract (urethra) from the bladder naturally narrows near the opening of the penis. If urinary crystals accumulate at this restricted passageway, they can form a plug that blocks the elimination of urine. As the bladder fills with urine that cannot be voided, the cat not only feels great physical discomfort, but is also in danger. Males are far more likely to suffer urinary blockages than are females whose urine passageway is slightly wider. Symptoms of feline urinary tract problems include excessive licking at the penis or vulva. The distressed cat may urinate outside the box in an unusual place such as the bathtub or sink, or may make frequent trips to the box and make prolonged attempts to pass urine. Some cats may dig in the box or vocalize excessively, hide or stop eating.

Urinary blockage is considered a medical emergency. If there is any doubt as to whether a pet is able to void urine, particularly if it is a male, immediate medical attention is advised.

10

Sex-Related Problems

Females should not be neutered before the first heat.

The recommended age for neutering female cats and dogs is around six months of age before the first estrus cycle or "heat." This advice is based on evidence that mammary tumors in dogs and cats are more frequent in animals that have had even one estrus. The incidence of mammary tumors (the equivalent of breast tumors in women) later in the pet's lifetime are far less likely to occur if the ovaries, which produce the tumor-stimulating hormone estrogen, are removed before estral cycling begins. The risk factor increases with every passing estral cycle which generally occurs every six months in the bitch and every three weeks in the queen! The risk jumps most dramatically after just one "heat."

Although not all mammary tumors in pets are malignant (cancerous), most pet owners will agree that it is not worth the risk. In cats, mammary tumors are more frequently malignant and more difficult to control than those in dogs. Because the first "heat" is frequently mild and may go unnoticed, the surgery should be scheduled according to your pet's individual rate of growth and physical maturity as directed by your attending veterinarian. Vaccination and general health status must be determined prior to any nonemergency hospital procedure.

A female dog or cat should have at least one "heat" or litter to be a better pet.

A female's basic temperament is unlikely to have any lasting influence from hormones that cycle during estrus or pregnancy. A pregnant female can

131

Never leave your bitch in heat unattended. Backyard fences are not barriers to determined males.

become more placid during pregnancy and lactation (while nursing the litter), but these have no enduring effect on the animal.

In fact, a mother can become more irritable and anxious while caring for a litter, and a female in heat is frequently more restless and overreactive. Intact or unneutered females are at increased risk of developing infections, problems related to hormonal "imbalances" such as cystic ovaries, and cancerous mammary tumors. Unless your pet is of outstanding show/breeding stock, she should be spayed. The "spay" (neutering) surgery involves removal of both ovaries and the uterus. This will minimize many medical problems during her lifetime.

Neutering your pet will eliminate the dramatic hormonal surges of the sexual cycle and will make her a more even-tempered pet in the long term while helping to control the pet overpopulation.

Male dogs or cats should not be neutered because sexual fulfillment is necessary for them to be happy.

Human beings have a perspective on life that is uniquely human. Our individual view of the world is a function of attitudes that are influenced by our family, neighborhood, town or city, nation, economic status, religion and times.

Sexual attitudes among people vary greatly. We alone have the distinction of reproduction as an option rather than an imperative. Human sexuality is not solely for the purpose of procreation. It is unreasonable to project our own exceptional attitudes regarding sexuality and sexual gratification onto our pets. These issues are only of any significance when real survival issues (like food, shelter, health) are secure.

Regardless of our personal opinions, one thing is for certain, we are not dogs or cats. Conversely, they are not human and will never share our perspectives on life. By their very definition as domesticated animals, pets are selectively bred (or not) under human supervision. The question of whether the act of intercourse is "enjoyable" to a nonhuman male or female remains open. Whether or not an individual of either gender "enjoys" the sex act, however, does not give him or her the right to reproduce.

There should be no double standard between neutering males and females. Any issues that delay or preclude a decision to neuter a pet must also include the morality of destroying millions of unneutered animals.

All bitches or queens instinctively know how to deliver and care for their young.

Compared to women, female dogs and cats are born with more complete reproductive instincts; however, this does not guarantee that their basic instincts are uniformly inherited. Although many of the behaviors governing the delivery and care of offspring are largely under the control of inborn behavioral mechanisms, these are not always strong or intact. A bitch can seem completely unaware of how to deal with the sudden appearance of pups still covered with a placenta and attached by umbilical cords. She may panic and try to run away or she may sever the cord too close to the pup's body. The maternal instinct seems to be much more consistent in cats, but there are always exceptions.

Instinct is not the only thing that governs a female's ability to care for her young, although it is obviously fundamental. Experience also contributes to the individual's quality of maternal care. The bitch that panics with the first pup, or even with the first litter, may improve with practice. Unfortunately, the cost of her learning may be high for the firstborn. Some cats and dogs unintentionally harm their pups by laying or stepping on them, or they may be negligent and rough. When undesirable maternal behavior is extreme in nature or does not improve with experience, the individual should not be bred again.

Injury inflicted on offspring by their dam is always accidental.

Kittens and puppies, like other baby animals, are vulnerable creatures. New-borns are completely dependent on their mothers for care, and this can be a critical time. Born deaf and blind, kittens and puppies rely on their mother for body heat, nourishment, hygiene and protection from any source of danger. This is a long list of responsibilities for any mother, but it is multiplied by the number in her litter and is magnified by her maternal experience and inclination. There is a lot of room for error.

Maternal instinct is the inherited portion that directs caring for the young. Learning from successive experiences with each puppy or kitten certainly contributes to the care a mother dog or cat will provide. However, no amount of learned experience will compensate for a mother who has inherited a set of extremely undesirable maternal qualities. Infanticide, or the killing of one's own progeny, is reported in pet dogs and, rarely, cats. This may be accidental in overzealous or overanxious mothers. Babies can be crushed or smothered to death. They can be eviscerated when mom cuts the umbilical cord or devoured during efforts to clean away the placenta. A mother may be overwhelmed with the care of a large litter and neglect some or all of her brood.

Maternal failure or incompetence is not always accidental. It should always be used, however, as a criterion for the individual's desirability for future breeding. A mother's distinctly unmotherly regard may reflect the health condition of her offspring. One or more newborns can be intentionally rejected when others are well cared for. Mother dogs and cats have a sense as to the health of their babies. In the interest of conserving their energy and concentrating it instead on offspring with higher survival potential, sick babies may be singled out as poor investments for maternal energy. This behavior may seem cruel at first, but it is Nature's way of ensuring the survival of those most likely to benefit.

Male cats or dogs recognize their own offspring and would never harm them.

In many species, the male may have no more to do with rearing the young beyond the act of mating. Cats and dogs require little or no paternal investment, and the mother of the litter usually bears the burdens of pregnancy and raising the litter alone.

Among domestic cats, there is a rare but real possibility that a tomcat will harm kittens. Although male infanticide is less of a risk in dogs, the female dog (bitch) often defends her offspring even from a familiar male and prevents his investigation regardless of his intent.

Even if he remains close by as the puppies or kittens mature, the male parent is unlikely to recognize his own offspring. Females may mate with multiple partners during sexual receptivity. Moreover, it is improbable that the male dog or cat makes any connection between the mating and any offspring that are produced approximately two months later. He can be predicted to interact with them as he would with any other animal. He will likely perceive male progeny as potential rivals, for example, and female offspring as potential mates.

Males without prior experience always know how to mate.

Instinct and learning by experience are important for successful mating. The learned component can be particularly important in male dogs. The initial attempts at mating by naive male dogs can be quite comical. They may literally mount the wrong end or approach the female from any number of awkward angles. With experience and a patient mate, they generally can be expected to get it right.

Although the tomcat's instincts regarding sexual behavior seem to be less influenced by external factors, he may be sensitive to the location of mating. Successful mating in cats is usually best if the tomcat is in familiar territory. This is true of many dogs as well and should always be considered when breeding is planned.

Allowing for individual differences, of course, environmental distractions can affect the male's drive. Both male dogs and cats are affected by the tolerance of their intended mate. If a female is uncooperative in any way, the male's efforts may be denied or at least delayed. For planned breeding, it is usually best to mate a virgin male with an experienced female. Of course, it would be helpful if a virgin female is bred to an experienced male, but it appears to be less of a concern.

Neutering will make a male or female dog less protective of you and your home.

When we refer to a dog's protectiveness, we are actually regarding several different types of defensive behavior:

- *Territorial behavior* can be seen in a dog's aggressiveness in protecting the owner's home, yard, car, neighborhood and any other area that has been previously investigated, patrolled or marked. Territorial behavior can be magnified by the value of specific locations, such as a favorite resting place. Protectiveness can also refer to the dog's inclination to defend the family.

- *Pack defense,* of people or of other dogs, is important in social animals that rely on each other's cooperation for survival. Protectiveness can also refer to the defense of valued objects. This behavior, known as *possessive* aggression ("guarding" behavior), can be quite severe even in animals that are not aggressive in any other situation. (See chapter 7, "Aggression," section entitled *Dogs will never bite the hand that feeds them.*)

Neutering is not expected to change your pet's "protectiveness." Territorial, pack-defensive or possessive aggressions are not determined by reproductive hormones. These behaviors are either part of individual genetic predisposition or they are not. Each of these patterns is then modified by life experiences and the reinforcement of behaviors in specific contexts. Spayed females, for instance, are not less territorially aggressive than intact (unspayed) females. Unless a behavior is due entirely to sex-related hormones, neutering the dog will not eliminate the behavior.

Neutering a male dog or cat is the same as a vasectomy performed on a man.

The vasectomy is not useful in veterinary medicine because the reasons for rendering a male dog or cat infertile are not the only concerns. Neutering, also called castration in the male, is the surgical removal of both testicles, rather than cutting off the supply of sperm, as with vasectomy. Neutering effectively prevents the birth of unwanted litters and takes care of many behavioral and medical concerns attributed to testosterone, the principle male sex hormone produced by the testicles.

Fighting, masturbation, roaming and inappropriate urination often decrease (but are not necessarily eliminated) by neutering the intact male. Testosterone-responsive tumors, among them tumors of the prostate, are significantly reduced or prevented by castration. Infections of the urogenital tract can be difficult if not impossible to eradicate even with the most appropriate antibiotics unless and until the dog is neutered.

Vasectomy alone simply does not apply in veterinary medicine. This surgery fails to accomplish the goals that castration meets. The male dog or cat is not embarrassed by the absence of testicles and will not suffer any type of social humiliation.

Concerned pet owners must be careful not to confuse what is best for their pet's long-term health and happiness with issues that matter only in terms of human sexuality.

A "spayed" female retains her ovaries, but the uterus is removed.

The procedure known as the "spay" is the layman's term for neutering a female dog or cat. This surgery is the equivalent of the hysterovariectomy in women. It involves surgical removal of both ovaries (when both are present) and the uterus. In human medicine, it is not always necessary or advisable to remove the ovaries if only the uterus is diseased. Removal of normal ovarian tissue is generally discouraged unless there is a family history of ovarian cancer or some other medical reason to remove ovaries. In women who suffer from diseases of the uterus, unless there is medical cause to remove the ovaries, only a hysterectomy (removal of the uterus alone) is performed.

There is no evidence that female dogs or cats suffer the emotional or physical symptoms of menopause familiar to many women. Hormone supplements are not prescribed for spayed pets because they do not have the same consequences routinely experienced by menopausal women. There is, therefore, little justification for performing a hysterectomy and not a complete hysterovariectomy ("spay") in pets.

In veterinary medicine, the "spay" is recommended to accomplish several goals:

- If only the uterus is removed and the ovaries remain (hysterectomy), the female will continue to cycle, to attract males and to be predisposed to a variety of undesirable behavioral and medical problems, some of which can be quite serious. Because the uterus is not a vital organ and serves no function without the ovaries, both ovaries and the uterus are normally removed.

- If only the ovaries are removed, diseases of the uterus may still occur and additional surgery, that might easily have been avoided, could be required later.

"False" pregnancy in a female is normal and indicates a desire to give birth.

"False" pregnancy is associated with hormonal fluctuations that persist unusually long after estrus. It occurs in both female dogs and cats, but it is far more common in the bitch. The affected female may appear to be pregnant, showing lactating mammary glands, an expanding abdomen and behavioral changes associated with preparation for delivery and even care of surrogate newborn.

In some cases early in the ovarian imbalance, male dogs may continue to be attracted to the female. This condition often resolves itself without treatment, but in the meantime it can be a stressful time for the patient and her owners. The female may be restless, anxious and possibly uncomfortable. It is not a "normal" condition and has nothing to do with the female's secret desire to reproduce. Animals that are not valuable breeding stock or those that have repeated or severe episodes of false pregnancy should be spayed.

Female cats will mate with any male cat, and female dogs will mate with any male dog.

Mate selection is recognized in cats and dogs but it is highly unpredictable. Still, it is important to acknowledge that it does occur. Some females will accept only certain males, and some males will refuse to mount just any female.

Mate preference, although still unusual, may be more common among dogs. For the most part, however, any individual preference will be overruled by strong reproductive drives. The biological motivation to maximize reproductive success is determined by species, availability of potential mates, mating opportunity, environmental conditions and individual factors such as health and mate preference. In people, these factors are further complicated by a variety of imposed social, cultural and religious directives. It is always a struggle between instinct and learning.

Dogs that have mated must be pulled apart or they will remain stuck together permanently.

When a male dog mates with a female dog, his penis swells near the base to occupy the width of her vagina. The effect of this canine peculiarity is to prevent the bitch from separating prior to ejaculation to increase the chances of a fertile mating. The disadvantage of this phenomenon is that both breeding partners remain locked in position for up to about half an hour until the swelling mechanism recedes. They are vulnerable to the unwelcome attention of intruders, including well-meaning but naive people.

Once they are locked together during the mating "tie," it is a mistake to try to separate breeding dogs. First, it would be exceedingly painful for both animals, and the effort would likely fail. Second, either or both animals could

The genital "lock" keeps dogs joined after mating (they often turn back to back), but they should not be disturbed until separation occurs naturally.

become aggressive toward the person trying to pry them apart because of pain or fear, or both. The male could become aggressively defensive of a mate. If you think you can prevent an unplanned mating and discover the pair already "stuck," you are already too late. Wait until they separate and contact your veterinarian that very day to discuss how to prevent unwanted pregnancies once mating has occurred. Although there are options available, the best way to prevent unwanted litters of puppies is to neuter male and female dogs.

It takes six months after neutering for a male's behavior to change.

After surgical removal of the testicles, the decline in testosterone, the male sex hormone, is immediate and irreversible. Castration eliminates the source and the levels of testosterone that are still in circulation plummet. Behaviors caused by testosterone can be expected to cease right away.

Most behaviors, however, result from many contributing physiological factors. If testosterone is only partially responsible for a given behavior, neutering will not prevent the behavior from recurring although it may be modified. Fighting between male cats, for example, will continue after castration, but will be less likely, and when it does occur, it will be less intense. Masturbation by neutered male dogs is considerably more likely to persist if the behavior began before they were neutered.

Although it is often tempting to pin the entire blame for undesirable behaviors on testosterone, sex-related hormones typically compose only part of a pet's predisposition toward a given behavior problem. If a behavior is predominantly under hormonal control, its modification and, perhaps, elimination can be expected to coincide with decreasing hormonal levels immediately after surgery.

Both males and females come into "heat" around six months of age.

The term "in heat" indicates that a nonhuman female animal is in estrus. Estrus is the period of reproductive receptivity and fertility in the estral cycle of many female mammals. Estrus, also referred to as "heat," coincides with the internal release of eggs from the ovaries for fertilization during mating.

The average female cat and dog reaches puberty around six months of age. This is not a fixed schedule, however, and individuals can occasionally begin cycling earlier or later. The first estrus is often a "silent heat," implying that the signs can be mild, passing unnoticed by owners, yet conspicuous to others of the same species. There is less variation in the onset of estrus in cats than in dogs, partly because of the greater range in size difference among dog breeds. In general, a bitch's predicted sexual maturity frequently occurs later than six months if she is large or giant size.

Male cats and dogs may attain sexual and reproductive maturity as early as six months as well. In dogs, this is partly influenced by their individual growth rate and the body size of the breed, since large breed dogs tend to mature later than smaller dogs. Males, however, do not "go into heat." The expressions "in heat" or "in season" apply to females only.

Under the influence of sexual hormones that surge at adolescence, the male (and female) can become restless and respond more determinedly to owner attempts to restrict its activities. The intact male dog and cat frequently attempt to escape outdoors to roam freely as puberty approaches and into adulthood. This does not mean the male is "in heat." He may, however, be responding to somebody else's "heat."

Female dogs menstruate just like women.

The onset of reproductive maturity among dogs is far more variable than among cats. Smaller breeds may begin cycling by five months of age compared to giant breeds that may not reach puberty until over one year of age. The first "heat" is expected between six months and one year for the average-sized dog. Normal estrus cycles in dogs last up to three weeks. There may be a mild to moderate amount of bloody discharge from the vagina in the first week or so. Ovulation, or the eruption of eggs from the ovaries, usually occurs between one to three days after the bitch accepts the stud. Receptivity precedes fertility and both occur roughly midway through estrus. The gestation period in dogs is about sixty-three days, give or take five days. Individuals may come into "heat" an average of twice annually, although some females cycle up to three or four times a year, and others cycle once yearly. In the bitch, vaginal bleeding precedes ovulation. In women, vaginal bleeding usually means a time of low fertility after ovulation is passed. Another prominent distinction between women and the rest of the animal world is that we remain sexually receptive independently of fertility. There is no true "breeding season" in people although there are seasonal fluctuations. Other species permit mating only for reproductive purposes and usually during certain times of the year.

The best way to tell if a bitch is in estrus is to watch for vaginal bleeding.

Vaginal bleeding does not always accompany a bitch's "heat." Some females have abundant bloody discharge preceding the fertile portion of estrus when the female will be receptive to the male's advances. Others show minimal vaginal discharge, if any. The amount of bleeding can also vary so that the same female may not always show heavy discharge. Very young or very old females may have relatively light discharge.

The best way to tell if a bitch is "in heat" is to examine her external genitalia and monitor her attractiveness and response to male dogs. A female in estrus will typically have a noticeably swollen vulva, the external skin that covers the vaginal opening.

If male dogs suddenly appear on your doorstep or make attempts to enter your yard or home, chances are that they are responding to chemical messages emitted from your bitch. Chemical signals in her urine and body odor signal that she is sexually "available." During the most fertile phase of estrus, the female will stand immobile when gentle hand pressure is applied to her lower back. The layperson's term for this phase is "standing heat." This implies that she should accept mounting by a courting male. Of course, if your bitch is in any phase of estrus, she should never be left unattended in your yard. The time of greatest receptivity is difficult to predict and can vary between individuals and from one cycle to another. You should also be prepared to fend off "suitors" during walks around the block.

It is important to closely supervise your pet to prevent unwanted attention from dogs that may result in unintended breeding or injury. Speak to your veterinarian about "spaying" (the term for neutering a female) your dog before her first heat. (See this chapter, section entitled *Females should not be neutered before the first heat.*) Seek immediate advice the same day that any undesired mating occurs.

Female cats come into estrus only once or twice a year.

The sexually mature female cat comes into heat about every two to three weeks. Estrus, or "heat," the period during which she will be sexually receptive and fertile, lasts about one week or so. This means that she will have only one or two weeks between cycling.

The breeding season is determined in part by the climate. In temperate zones, the season may last from January through November, although peak mating season is in spring and fall.

Ovulation is provoked by the act of mating (it occurs twenty-four to forty-eight hours after breeding) and does not occur without sexual contact with the male as it does in dogs and women. Protrusions cover the surface of the male's penis and irritate the vaginal mucosa during mating. A relay mechanism is triggered to inform the brain that mating has occurred and the brain, in turn, signals back to the ovaries to induce ovulation. This ensures that the release of eggs and mating always coincide to increase the chance of reproductive success. The gestation period for most pet cats is about sixty-one days, give or take three days.

The queen can begin cycling as early as four months or may not become sexually mature before the age of ten months. On average, queens become sexually mature close to six months. This is influenced by the season and by contact with other sexually mature cats. Females that are five months old in December, for example, and are kept indoors, may not have their first "heat" before March. Breed may also determine the age of estral onset. Persian cats tend to reach sexual maturity later compared to other breeds and may not cycle until they are over one year old.

Neutering will make a pet fat and lazy.

Neutering impacts primarily on behaviors which are under the strongest hormonal influence. The behavioral impact of the surgery is usually beneficial:

- Individuals mellow once they are neutered but in most cases general activity is altered little, if at all. Dramatic temperament change is not expected.

The surgery can also have physical consequences, the vast majority of which are distinctly positive:

- Prevention of serious infection and many forms of cancer make neutering worthwhile.

Weight gain is a possible undesirable consequence of neutering, however, it is not inevitable. Increased body fat can result from several things. Following surgery, neutered animals undergo minor metabolic changes that affect fat storage and can predispose the pet to gaining weight. In the immediate postoperative period, a pet is less active than usual. Activity may temporarily subside after any surgery or illness so that calories are stored rather than used. Consider also that most pets are neutered between the ages of six months to one year. At or not long after this age, playfulness may diminish for many individuals.

Modification in temperament or activity level associated with neutering may simply coincide with the pet's psychological maturity as it grows out of the juvenile phase. Sex-related hormones and their contribution to an animal's energies are deleted by neutering. Any change in overall activity is usually transitory, and the majority of pets return to normal within a short time. Neutering will not make a pet sluggish or stupid, but minor physiological and behavioral changes can become major weight gain if the owner is not careful. And therein lies the key.

The foremost cause of postsurgical weight gain is simply that many pet owners tend to overfeed their pets following surgery. Overfeeding can be intentional. Some pet owners overfeed recuperating pets because they feel guilty for causing their pets pain or are themselves in emotional conflict with regard to neutering their pet. Overfeeding can be unintentional. The most important way that pet owners contribute to obesity is that they do not adjust their pet's volume of food in keeping with reduced levels of activity. In other words, the owner must adjust and balance diet with exercise to maintain a pet's ideal weight.

To ensure a pet's form and fitness, whether it is neutered or not, a balance must be established between caloric intake in food and caloric output in activity. Control your pet's weight by feeding premeasured meals at regular intervals. Restrict access to food between meals. Daily exercise means walking your dog at a vigorous pace and for at least twenty minutes twice daily, if possible, to maximize aerobic benefit. Playtime is also important physical activity as is Obedience training. Play with your cat on invitation and by your own initiation.

If your pet, neutered or not, gets fat despite your efforts, then you must either decrease its food or increase its exercise. If exercise cannot be augmented further, then the food ration must be lowered or gradually switched to a lower calorie feed. Speak with your veterinarian about how to prevent and treat obesity in your pet throughout a lifetime.

Neutering will make a pet mean.

Aggressiveness of many types tends to decrease after neutering. The extent of their reduction is proportional to the component of hormonal control. Intermale aggression, for instance, may persist but it will likely be far less intense. The only type of aggression that might temporarily emerge after neutering is in the immediate postoperative period. Pain-induced irritability is common in patients recovering from an illness or surgery. In normal recuperation, however, any undesirable temperament change is short lived and subsides with healing. Fear-induced aggression can also occur for a time when the disoriented patient awakens from anesthesia or is briefly hospitalized. Irritability and fear can persist in the transitional first few days at home, but life soon returns to normal.

The long-term benefits of neutering a pet far outweigh any short-term discomfort that is part of any surgery.

Neutering will stop unwanted aggression in males and females.

Very few types of aggressive behavior come under the influence of sex hormones released either by ovaries or testicles. With few exceptions, such as fighting between male dogs or the maternal instinct to defend the young, the hormonal exacerbation of most aggressiveness is minimal. Hormones cannot be blamed for everything. Learning and opportunity can be even more important to a pet's tendency to behave in a certain way.

In dogs and cats, as in people, an individual's inborn temperament traits are shaped by experience as it interacts with the world. Pet owners must be aware of how their reactions shape the future actions of their pets. From an early age and from the moment a pet is introduced into our homes, dogs or cats observe our responses and learn how to solicit attention or food, or perhaps how to avoid contact or punishment. Be aware of how every action will determine your pet's reaction, immediately and for a lifetime. You are implicated in establishing either your pet's desirable patterns or others that might otherwise have been prevented. Neutering cannot undo what you have taught your pet to become or what you have allowed in behavior. If you expect that neutering your pet will cure undesirable behaviors, you will be setting yourself up for disappointment. Inquire into Obedience training or more specialized consultants as soon as any aggressive tendencies are noticed. Your aggressive pet should be neutered because it must not be allowed to perpetuate aggressive genes.

Neutering a male cat will create a predisposition to urinary blockages.

Surgical castration has no recognized association with a cat's individual predisposition to develop feline urinary tract disease (FUTD), including feline urological syndrome (FUS). There are many mechanisms, including diet content, bladder infection, exercise, water intake and obesity, that are recognized to contribute to FUTD. Neutering is not among them.

FUTD occurs in tomcats regardless of whether they have been neutered. Cat owners should not delay neutering for fear a pet could develop a urinary blockage. Discuss your concerns with your veterinarian who will guide you toward a preventative program for your healthy pet.

Neutering will prevent urine spraying in male cats.

(See chapter 9, "Elimination," section entitled *Urine marking is prevented by neutering a cat before six months of age.*)

Mounting behavior indicates a dog is sexually aroused

(See chapter 7, "Aggression," section entitled *Mounting behavior indicates a dog is sexually aroused.*)

Male dogs and cats do not have nipples.

Look again! All male and female mammals are born with nipples. In cats and dogs, nipples occur in two parallel rows along the underside. There are usually about three or four pairs of nipples, and hair surrounding each one may be sparse. In unspayed females, the nipples swell under the hormonal effects of estrus, pregnancy and lactation as the underlying mammary glands prepare to nurse the young. In males, nipples do not normally develop. Enlarged nipples in male animals can be a sign of medical disorders and should be reported. Males can also suffer from mammary tumors although these are far more common in females.

There are no sexually transmitted diseases in pets.

Sexually transmitted diseases occur in both cats and dogs. Many different bacterial infections can cause symptoms in both sexes with consequences that can range from mild to severe. Infections of the urogenital tract can result in minor discomfort, or they can cause the loss of pregnancies, irreversible infertility and even death. The consequences of sexually transmitted diseases, in pets as in people, are frequently more serious for the female than for the male.

One exception is the bacteria, *Brucella canis,* which induces a sexually transmitted disease in dogs that can equally impact either gender. In cats, viral diseases (two types of feline leukemia and feline infectious peritonitis, for example) can affect reproductive systems as well as the general health of the individual. It is unlikely that these are transmitted specifically through sexual contact, but are certainly transmitted at the time of sex.

In dogs, venereal tumors can become large and hemorrhagic, requiring surgical excision. Indiscriminate mating in cats and dogs should be prevented to minimize the health risk to the individual pet and to prevent unnecessary contributions to the pet overpopulation. Animals not intended for breeding should be neutered. Valuable breeding pets should be thoroughly examined by veterinarians. Prospective breeding pairs should be blood-tested for contagious feline (the feline leukemias, FIP) and canine diseases (Brucellosis) as well as genetic diseases prior to contact of any kind.

Males must be neutered early or they will learn to masturbate.

Self-stimulated sexual behavior is occasionally seen in male dogs and, rarely, in male cats. Masturbation is more likely to begin in individuals that have not been castrated. This is not to say that all males will masturbate unless they are neutered before a certain age. Most male cats or dogs, neutered or not, never perform this act. Some males that begin to masturbate before neutering will refrain afterward and masturbation can continue beyond castration.

An important and rarely mentioned contributing factor to masturbation in animals is the role played by observers. Initially, inappropriate self-stimulation can be comical. Onlookers may unintentionally (or intentionally) encourage the animal with their attention. Unfortunately, the very act of performing the behavior increases the likelihood of repetition. That is why early intervention in distracting the animal is so helpful. Observers should be careful not to confuse canine mounting, associated with social dominance, and mounting associated with sexual activity. Dominance-related mounting behavior can be displayed by both males and females and can be directed against another dog or even a person. Sexual mounting is typically directed by a male toward a receptive female in season.

Masturbation can be directed toward an inanimate surrogate object, such as a blanket, that can then become a habitual trigger for the animal. If a single target object is identified, early removal may be enough to stop the progression of the behavior. Sometimes this only instigates the individual to select another target. If your pet is engaging in behavior patterns of any kind which disturb you, consult your veterinarian for referral to a veterinary behaviorist in your area.

Dogs howl for many reasons—sometimes just because they can. (photo courtesy of Walter Grondell)

11

Miscellaneous Behavior Problems

Dogs howl because they are unhappy.

Dogs howl for many reasons, sometimes just because they can. Among breeds that howl a great deal are the Nordic dogs that, not coincidentally, most closely physically resemble their wolflike ancestors. Bloodhounds also howl, but produce a vocalization called a "bay," sounding like a bark-howl. Other dog breeds are not prone to howling at all.

Dogs howl for some of the same reasons that wolves are thought to howl. The plaintive wail is a vocal pattern that echoes for miles around, reaching other animals within earshot. The howl can be heard by pack members at a distance and also be a warning to wolves of other packs, reminding them of territorial boundaries. The pack howl, performed in groups of two or more, is performed in early morning and evening choruses. Howling in groups solidifies bonds between individuals and ensures pack unity. It is interesting to speculate on parallels of this behavior with the human practice of vocal choruses and singing in general. The howl of a solitary dog should be interpreted in context, perhaps separation from the owner or another companion. Some dogs howl when perfectly content in fulfillment of some ancient rhythm of life.

Dogs bark because they have something to say.

Barking is the most common vocal pattern in dogs. Terrier breeds tend to bark a lot. The exception is the Basenji, which does not *bark*, but emits other sounds. Interestingly, wolves rarely bark, and other vocalizations predominate. It is unclear whether the bark was intentionally emphasized in early selective breeding of the dog's wolflike ancestor or whether this was an incidental effect of domestication.

149

Barking is only partly determined by an inherited tendency. In addition to the obvious importance of genetic predisposition, learning to bark is an often underestimated element. Owners of new puppies often report on how quiet the young dog seems. Weeks later, they commonly exclaim that their puppy now seems to bark all the time. Owners unknowingly shape this behavior, and many others. When pups discover their voices and bark for the first time, they typically meet with the owner's enthusiastic attention and exclamation of delight. In a way, this resembles the parental response to a baby's first cooing, and later, the first word. If barking elicits immediate and clear rewards in the form of praise and attention, it is certain that the behavior will be attempted again. The pup will bark to elicit the same reaction. Barking typically becomes the most effective way a dog can get attention.

Dogs do not speak, although some dog owners teach their pets to bark on command using "speak" as the cue. Barking is not the same as language despite the wide range of tones, volume and apparent expression that modulate a dog's bark. Barking can be an aggressive vocalization when used by a dog defending territory against an intruder. A bark can be a soft "Woof?" as a dog enters the room in search of you. But in most cases, dogs bark because they can and because it is something to do, as a habitual form of daily expression. Research indicates that dogs may bark just because it is fun!

Pets must be sedated for travel.

It is natural to worry about a pet's comfort and well-being during travel. Our concern may be even greater if the pet is separated from us and confined in another compartment, such as in the baggage section of an aircraft. In the majority of cases, a dog or cat will be much better off if it is *not* medicated during travel. For one thing, sedatives have side effects and can cause changes in blood pressure or heart rhythm. Your pet may be even more anxious because the tranquilizer may make the animal feel strange or disoriented. Even if your pet experiences some fear during travel, the ability to regulate body temperature and to react to jarring motions will be better if the animal's body is left to its own devices. A healthy pet can cope with periods of stress and will soon forget any anxious moments when you appear at the end of the trip.

If your pet has any serious health concerns, sedatives may not be advisable. In fact, anything less than a healthy pet should probably not travel unless it is absolutely necessary. If your pet panics inside a crate thereby risking injury, sedation might be of benefit. Ask your veterinarian to prescribe an appropriate medication, but do not rely on the medication to ease your pet's transport.

If you are planning a trip, it might be a good idea to give the medication to your pet before you go anywhere so that you can see how the animal responds. Your cat or dog can associate the unfamiliar effects of sedatives with the security of a familiar environment and might adjust better to the sensation of being drowsy. Before going on long excursions, take frequent short trips in your car to get your pet more familiar with being transported. Allow your pet to become accustomed to being crated for a few minutes every day so that the crate becomes a familiar and secure place.

Crate training is the best way to raise a dog.

Raising a pet is much like raising a child. Each one is an individual. General guidelines and helpful hints can be of benefit to both you and your pet, but not every recommended training method will apply to every individual.

Crate training requires the confinement of a dog to a cage to restrict activity when the owner is unavailable. Although crating a pet can successfully prevent house soiling and destructiveness in the owner's absence, it is not for every dog. Some dogs panic when confined in such a small space and will do anything to get out, even though the crate may be of appropriate size. This is usually caused by improper introduction and use of the crate. Dogs have chewed, clawed and bulldozed their way through seemingly indestructible cages, frequently injuring themselves in the process.

Many dogs will become accustomed to the crate over time and voluntarily return to have a nap or enjoy a favorite chew toy. Almost all puppies will exhibit some anxiety when first confined. If a puppy cries and barks when placed in the crate to get attention, it may be too much for some dog owners to ignore. Once a puppy knows that vocalizing will eventually lead to liberation from captivity, vocalizing will persist. In other words, if you have decided to give crate training a try, do not give in to your puppy's initial protests, or it will never work.

On the other hand, do not feel that it must work at the expense of your puppy's emotional well-being. There are other less restrictive methods for various aspects of dog raising. Crate training is not house training your dog. Feed your pup at regular intervals and go for walks immediately after feeding. Dogs do indeed void in their crate if they lack the training or opportunity to eliminate in appropriate locations.

One risk of this method is that pups and dogs will remain confined when their owners are at home but do not want to be bothered with them. Owners frequently misunderstand the purpose of a crate and leave the puppy isolated there for extended periods. *The truth is that dogs of any age need social contact with their owners more than anything.* Deprived of opportunities to

Critical to successful crate training, the choice of crate must be comfortable for your pet's current and predicted adult size. A crate too large or too small defeats the purpose.

socialize during a critical phase in their development, pups may become maladjusted to social interaction with people and even other pets.

If you have decided to try crating your dog, do so gradually for brief periods and reward good behavior. The dog should not be crated unless you are away and if you must leave, make sure your absences are short. If you want to try confinement overnight, be certain your dog is tired and satisfied in every way before you retire to your own bed.

Many owners put the dog in a crate as a form of punishment and then wonder why the dog becomes anxious or refuses to enter the crate altogether. Used with common sense and sensitivity, crate training can be very effective for many dogs. It is not the universal answer, however, for every problem or every dog owner.

A dog that rolls over in greeting has been abused.

Rolling over is the most submissive posture a dog can display. It is an intentionally exaggerated display so that there is no social ambiguity.

Dogs assume a vulnerable position in the presence of another dog or person that is socially dominant over them. In a submissive roll, the dog avoids direct eye contact and looks away, thereby exposing its neck to the dominating individual. This extreme form of ritualized submission is seen in very young dogs or in socially anxious situations. Some signs of submission are often more subtle. The submissive dog will avert its eyes, tolerate mounting or simply sit down.

An animal that has been abused or traumatized in a given situation or by a specific individual may anxiously roll over in a submissive greeting. Animals that have never been abused can do this, too. *The submissive roll is not a reliable indicator of past abuse.* It is merely an indication that your pet is insecure in the current situation and may simply testify to your pet's acknowledgment of a socially inferior status relative to another person or animal, based perhaps on a previous encounter.

Hyperactive pups require sedatives or psychoactive drugs.

Pathological hyperactivity, referred to as hyperkinesis, is rare in dogs and can require specialized intervention. Overactivity can also be a sign of a medical disorder. In cats, for example, hyperactive thyroid glands are commonly related to an increase in normal levels of activity. Female dogs or cats in heat can behave restlessly.

Murray, a Husky crossbreed, in a submissive roll. (photo courtesy of Carolyn and James Tanner)

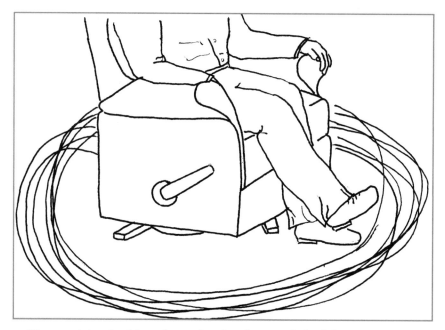

Hyperactivity should not be confused with normal playful activity.

Young healthy pets are normally active. Normal playful behavior can take a variety of forms. It can range from curious investigation of every corner in your home to mad dashes and frenzied racing around. Sudden bursts of energetic play are usually short lived compared to less frenetic and more extended periods of mischief. *Neither form of play is a symptom of hyperactivity, however, and should certainly not be suppressed with any medication.*

Playful energies should be directed toward acceptable outlets of activity as much as possible. Puppies should be walked outdoors at least several times every day. Appropriate games should direct attention toward balls and other toys so that targets never become parts of a person's body or clothing. Obedience training should begin from the moment you acquire your pup so that the training can become an enjoyable and constructive interaction. Consult your veterinarian if you feel your pet's energy is unusual or excessive.

Pets eventually outgrow a fear of thunder or other loud noises.

Extreme forms of fear that seem out of proportion to the actual threat imposed are referred to as phobias. Phobic responses can be acquired after a single terrifying exposure to the vacuum cleaner, for instance, or can progress more gradually over successive exposures, such as yearly visits to the veterinary clinic. Fear can be associated with a strange context or unfamiliar person. Fear can also be instinctive in certain circumstances—loud and sudden noises naturally startle a dog or cat. If the noise is particularly powerful, like the sound of a truck backfiring, instinctive defenses drive the pet to seek shelter or escape any way possible. Phobic dogs have been reported to leap through windows, tear screen doors and to generally jump over or chew and scratch through any barrier that blocks their path to a safe place. A phobic cat is truly a force to be reckoned with if pushed into fear-induced aggression.

Sedatives and tranquilizers may provide temporary relief, but will not resolve phobias in the long run. In fact, medication is often of little benefit, and their only effect may be to slow the pet's efforts to run or hide. In most cases where a dog or cat has developed an exaggerated fearful reaction to loud stimuli (thunderstorms, vacuum cleaners, traffic, fireworks), the fear can remain constant or increase over time. Phobic responses are unlikely to be resolved without intervention.

Although progressive retraining may seem tedious, it is the most effective way to counteract many phobias. Specific training combines gradual exposure to low intensity sound, for instance, with reward for calm behavior. For example, recordings of thunderstorms can be played at very low volume while playing with your pet or feeding a special treat. After many practice sessions over many weeks, the volume is gradually increased until the pet no longer is bothered by the sound of thunder even at realistic levels of playback. A combination of behavior modification with appropriate medication in a few cases may be necessary. Ask your veterinarian for a referral to a veterinary behaviorist for appropriate advice.

A pet that is fearful of men has been beaten by one.

Fear is an instinctive emotional and physical reaction. It is based in ancestral defense mechanisms triggered in the face of novel or unfamiliar situations. Fear responses can also be learned, after just one exposure to a menacing context or after repeated negative experiences, and can become a predictable part of the individual's behavior.

In a time when we are all keenly aware of the abuse of the potentially vulnerable and defenseless, we are quick to suspect that a pet's fear of an individual must be based in actual experience. In fact, most pets fear particular individuals as a result of inexperience. Social confidence stems from the exposure of young animals to a variety of individuals of different ages, genders and even races.

Pets that are raised in a household of women where male visitors are infrequent, rarely get the opportunity to socialize with men. There is a window of emotional development (generally thought to be between six and thirteen weeks for puppies, and it may be even briefer in kittens) during which any and all encounters with people will shape behavior for future interaction. A dog or cat that has limited experience with men or women, may react fearfully at first encounter with a stranger. Subsequent meetings may amplify anxiety if meeting this man or woman is not made a clearly positive event.

Once you recognize your pet's anxiety in certain situations, you may begin to gradually expose your pet to the perceived menace over successive controlled meetings:

- Ask the person to stop at a distance that is easily tolerated by your pet.

- The person should then approach your pet very slowly and stop at a slightly uncomfortable distance, going a bit closer at each visit.

- Loud voices or sudden motion are not helpful during these initial encounters.

- Direct eye contact should be minimized.

- Reward your pet's calm behavior with a special treat.

- Eventually, it can be helpful to ask the individual to participate in feeding your pet.

Once the pet seems more relaxed, ask your visitor to invite your dog or cat to play a favorite game. Give this new relationship a chance to grow over many brief and controlled visits. You would not want to overwhelm your pet's capacity to accept a new relationship or situation.

Pets that investigate the "private parts" of other dogs and people are abnormal.

All animals, and that includes people, have individual odors. Some of the most characteristic and recognizable odors emanate from the genitalia and surrounding region. Odors which distinguish each individual can be temporarily hidden by bathing or perfume but they are, nonetheless, normal and natural in every healthy animal.

Dogs and cats learn to identify each other by facial recognition and the detection of familiar scent. Cats cannot speak to each other to remind

Social greeting between dogs normally includes facial (above) and anogenital investigation (left).

themselves of where they originally were introduced. Dogs cannot call each other by name. Anogenital investigation is an effective way for them to meet and greet each other. This behavior is direct, honest and uninhibited.

Some people are offended when they observe this greeting or are embarrassed when a dog unabashedly investigates them in the same manner. Human beings persist in the fantasy that we are somehow above any "animal behavior" and deny our own links to nature. Anogenital investigation between pets is not abnormal. When directed toward a person it does not mean that person needs to bathe. One of the gifts that our pets bring to us is a renewed bond with nature and insight into our own natures. With them, at least, we too can be direct, honest and uninhibited.

Jumping on people is a dog's way of showing affection.

Canine affection can be displayed in a variety of subtle and not so subtle ways. It can shine in a gentle gaze as a dog observes you from across the room. It can also be the way a dog comes charging down the hall flying toward you. Jumping can be a nonspecific response to excitement or stimulating situations. Many behaviors, however, have more than one meaning and jumping on people is one of them. For example, jumping in the context of greeting a family member can be an expression of joy. A dog may jump on the owner in an effort to communicate anxiety due to fear or the discomfort of a full bladder or bowel. Jumping is also one of the ways that dogs assert their dominance over another individual. Jumping raises the dog to a superior physical posture by placing front paws as high as they can reach. We tend to interpret our dog's actions as we would another person's and, consequently, the dog usually gets praised for this aggressive act. We rarely stop to consider the subtle social implication of this behavior.

Not every jumping dog necessarily intends to socially challenge anyone. But because these intentions can coincide during seemingly innocent greetings, it is wiser to cautiously prevent any misunderstanding. Teach your dog to greet you in a more "civilized" manner. You can greet each other, still showing your affection and shared enthusiasm at your reuniting, without encouraging chaotic and inappropriate behavior. Instead, be calm and speak in soothing tones. Insist that your dog sit and stay so that you remain in control and do not signal a willingness to change the balance of power in favor of your dog. Your dog will be no less happy to see you.

Cat toys are only for cats, and dog toys are only for dogs.

The truth is that pet toy manufacturers target their toys to attract cat or dog owners. The average pet owner knows what toys will best entertain their own pets. Not all dogs are interested in balls and not every cat will be instantly drawn to a stuffed mouse, with or without catnip.

For cats in particular, the greatest success can sometimes be found with toys made at home from common items such as paper, string and aluminum foil. Small dog breeds may be very happy to play with some of the items marketed for cats. Some cats will tirelessly fetch and retrieve large toys meant for big dogs. Make sure that the toy you are about to purchase is appropriate for your pet. Check for removable parts that are easily swallowed. Strict regulations govern the production of toys for children, but no such supervision over pet toy manufacturing currently exists. If you have young children in addition to pets, make very sure that any pet toy you purchase protects your child as well as your pet.

Dogs and cats will outgrow car sickness.

The primary cause of motion sickness is a difficulty in balancing sensory information from the inner ear with the sight of objects passing by at high speed.

Car sickness is aggravated by emotions that make travel a nauseating experience. Excitement and anticipation mix with anxiety or fear to increase heart and respiration rate. Air can be taken into the digestive tract and lead to abdominal distension. If the pet only goes for a ride to the veterinary clinic or boarding kennel, the animals will learn that car travel is a negative event and may never "outgrow" motion sickness. Nausea could become a conditioned response (much like Pavlov's dog salivated at the sound of a bell) to car rides.

The best way to teach your pet to relax in the car is gradual exposure to separate elements of car travel and then to gradually longer trips. If your pet is always transported in a crate, for instance, that is a good place to start. Instead of storing the crate away, feed your pet in it and store favorite toys there. Next, bring your pet into the car (in or out of the crate), and sit in your driveway for a while. Talking to your pet, as well as petting, brushing or giving a treat is soothing. Then take the animal back inside.

Your progress should be slow so that your pet is completely comfortable before moving on. The next phase is to just start the engine and stay "in park" (windows slightly open, of course). Finally, go for frequent short trips around the block, then around the neighborhood, then around town. You might even want to pass by the veterinary clinic without going in. Avoid traveling immediately after a meal. Wait at least an hour or two, giving your pet the opportunity to empty bowel and bladder before travel.

If your pet still becomes nauseated, ask your veterinarian to prescribe medication that specifically combats nausea rather than any potent sedatives. Medication, if necessary, will complement your retraining efforts, *but should not replace them.* In time, most pets become accustomed to the sensation of motion and accompany us more confidently wherever we go.

Pets should never be boarded because they will stop eating and might die.

In an ideal world, we would be able to take our pets with us wherever we go. In case you were wondering, you have the right to live your life and to enjoy

Pets do not have a concept of time as we do and will be more affected by the quality of their care at a kennel than by the length of your absence.

yourself. You must not fill yourself with guilt and allow pet ownership to turn your home into a prison. Sometimes it is necessary to get away from it all, including your pet. An emergency could arise that gives you little choice.

Most pets adjust well to being lodged in a kennel. They may eat less well for the first couple of days, but soon fall into the rhythm of feeding schedules and cleaning routines, becoming familiar with animal keepers and confinement. Change is stressful. Occasionally, the stress of boarding might activate hidden illnesses in your pet. Fortunately, the tragic cases of pets that have completely refused food or suddenly become ill are exceptional and exceedingly rare.

Have your pet thoroughly examined and all vaccines updated by your veterinarian shortly before your planned departure. Choose your boarding kennel carefully and ask them how they respond to a pet that refuses to eat after more than two days. Give them the name of your veterinarian. If possible, practice boarding your pet for brief overnight stays or even a weekend before any extended absence, so that the kennel becomes a familiar place.

Many pets actually seem to enjoy their kennels almost the way many children enjoy summer camp. Remember that people are the only creatures that can tell time. Your pet is unlikely to distinguish between two days or two weeks. Although it is true that guilt makes the world go 'round, go ahead and have a good time. When you return, your pet will only care how good it is to have you home.

Pets should never be kenneled longer than one or two weeks because they will think they have been abandoned.

The only animal that can tell time is the human animal. Clocks and watches are human inventions. The intervals between sunrise and sunset, between full moon and new moon, were divided into seconds, minutes and hours by people. The rest of the animal kingdom responds unconsciously to daily and seasonal rhythms. Separated from their owners, a dog or cat is unlikely to know exactly how long owners have been away. Only people punish you with grudges for leaving them. All that really counts to your pet is that you come back, safe and sound.

Cats do better during separations because they do not love their owners.

Dogs are instinctively pack animals. It is their fundamental social nature to seek out the company of their own kind or the closest substitute. Cats are a different species with a different kind of intelligence and sensitivity.

Cats are naturally more introverted than dogs. Their displays of affection are frequently more subtle and brief but no less genuine. The feline talent to entertain itself may be a kind of independence, but this is not unfeeling. Cats, like dogs, are individuals. Some adjust well to their owner's absence and others do less well. During periods of social isolation, a dog may frantically chew or dig at the door, whereas a cat may quietly but no less excessively groom itself to the point of creating bald spots. Those who proclaim that cats do not form strong bonds with other cats or people simply do not know cats. It is their loss.

Choose the boarding facility with care. Make sure to leave the telephone number of your veterinarian in case it is required. Have all vaccines boosted and your pet's health verified by your veterinarian before you leave. Professional pet-sitters (or friends and relatives) may be better for some cats, so that they would have the comforts of home and be surrounded by familiar sights and sounds until you come home.

All boarding kennels are alike.

Choosing a boarding kennel for your pet should be as important as choosing your child's daycare or school. Don't just rely on word of mouth for either. Make your own assessment firsthand by visiting the place and people that you trust with the care of your pet.

You could drop by unannounced and ask for a tour, but many places would appreciate or even require a phone call to let them know you'd like to come by. If your request is denied, go somewhere else regardless of their reputation or how convenient the location or competitive the rates.

When you visit, speak with any and all staff members you see. Ask them if they enjoy their work and how long they've been employed there. Ask them if they would board their own pet at the facility. Find out every detail about feeding schedules, exercise routines, how much attention your pet can expect from the employees and early or special pickup times. Consider how clean the cages look and how clean the kennel smells. Notice if it feels damp, cold, overheated or stuffy. Note the appearance of other pets.

There are many things that go into making a boarding kennel good or bad or somewhere in between. The staff of people, from management level to kennel keepers, are critical to your pet's care. Make them aware of your pet's special needs and always leave a phone number in case of an emergency. If you will be difficult to reach, call in to inquire or delegate a friend to assume responsibility in your absence. Give your designated representative a signed letter of authority and confirm this with kennel personnel.

Be choosy about the boarding facility, but your expectations should be reasonable. No one will be able to provide the comfort or familiarity of your home. Busy kennel workers cannot give the time that you do or pretend the same affection when interacting with your own dog or cat. What you should be watching for is a generally clean and efficiently run place and a sense of decency about the people who work there. If you feel uneasy about anything you see, voice your concerns. If the response is unsatisfactory or you simply do not feel comfortable, choose another kennel or make other arrangements. Your pet would thank you.

12

Health

Pets do not suffer from the same diseases as people.

Dogs and cats can develop the same or similar diseases that afflict people. They can also develop medical problems that do not affect us.

The vast majority of diseases that can affect our pets are not contagious to us. Infections or inflammations of the eye and ear are generally not contagious to people. Most viral diseases affect a specific species. There are medical problems, some creating mild discomfort and others with more severe potential, that are transmitted between species. Notable and common examples are superficial fungal infections ("ringworm") and several kinds of intestinal parasites. Diseases such as rabies can also be transmitted by dogs or cats that have been exposed to other rabid animals.

Current advances in veterinary medicine have made it possible to diagnose and treat disorders that include benign or malignant tumors of the skin, brain, thyroid gland and digestive tract or anywhere else; hay fever and a wide variety of allergic problems; torn ligaments, herniated "slipped" disks of the spinal column, fractured bones, arthritis of many types; cataracts, conjunctivitis, glaucoma; sinusitis, tonsillitis and pneumonia. Diseases can be caused by viruses, funguses and bacteria. They can be secondary to metabolic and immune-mediated disorders. Medical problems can be traumatic (auto accidents, etc.) and/or genetic (hemophilia) in origin. Periodontal and plastic surgery, ultrasound and CAT scan, cardiac pacemakers and open-heart surgery are all within veterinary technology. Of course, we all hope you will never have a need to discover any of these, but they are available if you do.

Medication prescribed for people may be used for pets with the same symptoms.

There are a great many parallels in the anatomy and physiology of all mammals. Consequently, they can be diagnosed with similar or identical medical disorders. However, any medication prescribed for a sick pet may not be the same for a related disease in people. There are important differences between species regarding the tolerance, dosage and application of therapeutic drugs. Acetaminophen, for example, is very toxic in cats. Even aspirin, if prescribed for a cat, must be administered in minute quantities at intervals of every seventy-two hours. Antibiotics frequently require higher doses in dogs and cats than in people to reach the desired blood levels. Some drugs used to fight infection cannot be used in young kittens or puppies.

It is never a good idea to administer any medication or home remedy to your pet without first consulting a veterinarian who knows you both. Not only can you harm your pet, but you could unintentionally give a medication or dosage of drug that might make things worse. You might also interfere with the diagnosis or treatment of the problem. *Any pet in a medical crisis should be seen at an emergency clinic.*

Unless previously directed by your pet's doctor, resist the urge to treat your own animal. If your pet is allergic to penicillin, for example, you might not know that amoxicillin is a related antibiotic that will trigger the same fatal reaction. Unless your veterinarian is very familiar with your pet's medical history, you must have your pet examined before any medication is prescribed. A few minutes of your time could make a lifetime of difference for your pet.

Animals heal themselves by licking.

It is a natural response for many animals to lick their wounds. This behavior indeed has survival value, or it would not be so widespread. Even among human beings, it is instinctive to place a sore finger in the mouth. The action of licking an injured area is soothing partly because it interrupts the perception of pain. It also stimulates local blood circulation, thereby speeding the arrival of defensive white blood cells that act against infection and begin wound repair. The saliva contains enzymes which minimize bacterial invasion into broken skin surfaces.

In some cases, however, licking can complicate wound healing. Oral bacteria contained in the saliva can occasionally infect a wound licked elsewhere on the body. Excessive licking can prevent or delay wound closure. In other cases, licking is just not enough. Aggressive bacterial infections are not always contained and broken bones may not heal without veterinary intervention. If your pets are ill or injured, do not assume they can heal themselves.

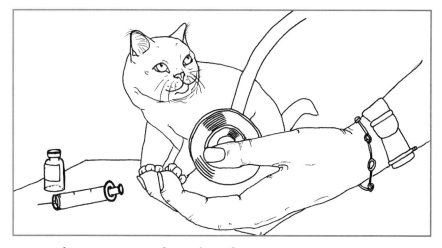

*Annual veterinary visits begin from the moment you acquire your pet,
regardless of age, and they continue for a lifetime.*

For a small abrasion or cut with minimal blood loss, clean it right away
with cool water and soap and rinse with clean water or hydrogen peroxide.
For more serious lacerations, apply direct and firm pressure to the site of blood
loss and bring your pet immediately to an emergency facility. Any unusual
odor or ongoing drainage should be reported to your veterinarian, even if the
wound itself seems minor or your pet is otherwise unaffected.

A warm, dry nose means your pet is ill.

There are many clues that indicate when dogs or cats are not feeling well.
Unfortunately, feeling the nose is not a reliable way to evaluate your pet's health.

The normal body temperature of cats and dogs is generally between 100
and 103 degrees Fahrenheit (about 37.5 to 39.5 degrees Celsius). Their body
temperatures are normally a bit higher than ours, so it is not surprising that
they feel warmer to our touch even when they are well. A dog or cat in a more
inactive or passive state, such as a nap, will have a nose that feels warm and
dry to your touch. On the other hand, the animal could be inactive and
lethargic because of feeling ill.

A pet can have a warm nose and be either in perfectly good health or
truly ill. Many signs other than the nose are more reliable reflections of your
pet's well-being. It makes better sense to assess appetite and interest in
pursuing normal activities or on obvious signs of a problem like diarrhea or
lameness. You know what is normal for your pet better than anyone. Rely on
your instincts and seek veterinary care even if you have any doubt.

A cold and wet nose means your pet is healthy,

The coldness or dampness of a dog's or cat's nose is really more an indication of activity level. When pets are resting or sleeping, their noses may feel warmer and even a bit dry. When exercising, playing or engaging in any other pursuit, the animal's nose will be cold and wet. Do not use your pet's nose as the only means of assessing illness. Many medical conditions do not affect the feel of the nose. The coldness and wetness of your pet's nose, however, will not provide any useful insight into health or even body temperature. That's what rectal thermometers are for!

Trust your judgment of your pet's health. If you have a feeling that something somehow is different, no matter how vague, or that the animal just doesn't seem to be completely well, you are probably right. Your veterinarian will want to hear about your suspicions and observations.

Vaccination is important only if your pet goes outside or has direct contact with other animals.

Vaccines boost your pet's immune system by stimulating the body's own ability to protect against specific viral or bacterial diseases. Because our pets usually do not produce a steady stream of antibodies against specific diseases, these vaccines must be repeated annually for life. Even pets that have little contact with other animals or those that remain indoors are still open to infectious agents that are transported, for example, in air currents.

The rabies vaccine, for instance, is now required by law in some states whether your cat or dog goes outside or not. In addition to the vaccinations, you will have regular opportunity to discuss any of your concerns with your pet's doctor. Ongoing problems can be monitored, and others you may not have noticed can be detected. There are many things against which we cannot protect our loved ones. It just makes sense to do whatever is currently available to ensure their health and safety. Most vaccines provide adequate defense against contagious diseases, but your pet could still contract a disease against which it has been inoculated. Vaccines may not be perfect, but the symptoms and consequences of an illness often will be less serious because of them.

Vaccinations guarantee your pet will never get sick.

Although most vaccines are researched and tested by the manufacturer, no vaccine is perfect because we live in an imperfect world. The response to the vaccine depends, in part, on your pet's immune system. If your pet is already incubating a viral infection or is in some other state that compromises the immune response at the time of vaccination, the vaccine will likely fail to protect it as well as possible. Even if your pet is inoculated against a specific virus, for example, it could still become ill following contact with the virus. Viruses have the ability to modify themselves slightly by the process known as mutation and overcome the protection afforded by the manufactured vaccine. Still, your pet might have been even more ill had it not been given any defensive boost from a vaccine.

Another problem associated with vaccination is the very small possibility of reaction to the vaccine itself. Sometimes what a pet owner calls a "reaction" is really due to local inflammation that can be an unavoidable part of giving any injection. Limping, for example, may be noticed for several days after an intramuscular rabies vaccine is given. True vaccine "reactions" appear soon after the injection. In dogs and cats, these can be mild to severe itchiness or progress to facial swelling and a difficulty in breathing. Diarrhea and vomiting can also be due to hypersensitivity to a vaccine. These must be immediately reported to your veterinarian so that treatment can be rapid and any necessary precautions can be taken the next time vaccines are due. Unless otherwise advised, it remains essential to vaccinate your pet at regular intervals according to your veterinarian's recommendations. The potential for benefit far outweighs any disadvantage.

The distemper vaccine prevents dogs and cats from biting.

There is no relationship between the vaccine generally referred to as the "distemper" vaccine and a pet's temperament. The vaccine commonly called "distemper" for cats as well as for dogs actually protects them against distinct sets of viral diseases that affect only cats or only dogs. Neither set of vaccines has anything to do with whether a cat or dog will bite or become aggressive.

For dogs and cats, there are two basic vaccines that should be repeated yearly or as directed by your veterinarian. The feline "distemper" vaccine inoculates the cat against several common viruses which affect the upper respiratory system. It also protects against another viral disease called panleucopenia, that can be more serious and affect many body systems including the brain. The canine distemper vaccine actually does trigger a dog's immunity against a disease that is called "distemper." This virus causes an illness that can severely affect the nervous, digestive and respiratory systems. The distemper vaccine for dogs also protects them against other diseases that cause gastroenteritis and bronchopneumonia. The feline viruses do not affect dogs and vice versa.

The notable exception is rabies, which can be transmitted through contact with blood or saliva of a rabid animal to any other warm-blooded animal (all mammals are susceptible, including people). The rabies vaccine, which must also be repeated at regular intervals, is the same for both cats and dogs. It protects them from developing the disease should they ever have the misfortune of exposure to a rabid animal.

Intestinal parasites affect only young pets.

Parasitic worms in the digestive tract are usually (but not always) orally transmitted in the form of microscopic eggs. This can occur from direct contact between individuals, such as between a mother animal and her offspring or between animals when they groom or lick each other in greeting. The transmission of parasite eggs can also occur from indirect contact. If a dog infested with worms defecates where other dogs play, their contact with this stool can perpetuate the parasite's life cycle. Even if the waste deposit is long washed away, the eggs can remain and often survive extreme weather conditions. Your pet could step onto the microscopic eggs and ingest them later by self-grooming.

Pets can become infested by intestinal parasites at any age. Even pets that never leave your home can contract worms. You might unknowingly carry the eggs on the soles of your shoes and deliver them into your home. If dogs or cats catch and swallow flies from the feces of an infested animal miles away, they could also become parasitized. If your dog never leaves your yard, other animals may still visit your property to deposit their solid waste as a way of marking their territory. Some dogs develop the unfortunate habit of eating the stools of other animals and become infested in this way. Dogs that ingest their own waste could become reinfested even after treatment for worms.

The transmission of intestinal parasites is unrelated to age. Stool samples should be analyzed at least once or twice every year to detect the presence of parasites before they make your pet ill.

Intestinal parasites always cause diarrhea and vomiting.

In many cases, the presence of intestinal parasites is tolerated by the "host" animal. The worms thrive, robbing the infested dog or cat of nutrients. Meanwhile eggs are released through the feces to extend the cycle of infestation to other potential hosts. The animal may not have any signs of gastrointestinal inflammation and may not even appear sick. It can even come as a surprise to pet owners when they are informed that their apparently healthy pets had parasitic eggs in their stools.

Sometimes, intestinal parasites cause symptoms that are vague and seemingly unrelated to the digestive system. A parasitized animal may become anemic or less resistant to infection, and have a dry, dull coat or itchy, flaky skin. The larval stages of some internal parasites travel through the lungs to provoke intermittent moist coughing. Weight loss can be dramatic despite a healthy or augmented appetite. Other individuals simply become depressed, eating less or not at all as they withdraw from normal social interaction. By the time any of these symptoms appear, diarrhea or vomiting may well commence, but even these effects may never be seen in heavily parasitized pets.

Parasitic worms can always be identified by looking at them.

With few exceptions, most adult worms look alike to the naked eye. Round-worm looks like long spaghetti noodles. However, in the larval stage the roundworm can be much shorter and resemble other worms. The exception is the tapeworm, the eggs of which infrequently appear in fecal analyses, which sheds clearly visible flat segments that resemble white rice and adhere to your pet's rear end.

The identification of parasites is necessary to determine the medication for effective treatment. It is also important in order to advise you when and if the treatment should be repeated according to the particular life cycle of each type of worm. This is also helpful in giving hints to avoid reinfestation. In preparation for microscopic examination, your pet's stool is liquefied and filtered. A sample is then scanned under the microscope for the presence of eggs that are typical of each and every parasite. All canine whip worm eggs look alike, but appear very different compared to coccidia, for example, another type of intestinal parasite.

Dogs should have a stool sample collected in the early spring and again in the fall to screen for parasites that may otherwise go undetected. An animal can be heavily parasitized and still present a batch of stool that contains no eggs. This is due to the cycle in which the female worms lay their eggs. In many instances where intestinal parasites are strongly suspected but have not been found in a stool sample, veterinarians will recommend worming (treating for worms) the animal anyway, just in case. If your cat roams outdoors and rarely uses a litter box, a fecal sample may be difficult to obtain. Even house cats benefit from fecal analyses since, as mentioned previously, it is still possible to bring eggs home on your own shoes or for your cat to swallow an insect harboring an egg from a parasitic worm.

Heartworm is detected by analyzing stool samples.

The adult heartworm is a parasitic worm that inhabits the heart chamber. It is not found in the intestine and is not eliminated in feces.

To test for heartworm, your veterinarian will recommend an annual or semiannual blood test. There are several different methods of analyzing the blood sample. Prevention is even more important for heartworm than for intestinal parasites because the consequences of heartworm infestation are so much more life threatening. *Medication to prevent heartworm disease is available through veterinarians but can only be administered to dogs that have tested to be free of heartworm. A dog with heartworm that is given the heartworm preventative could have a fatal reaction.*

Heartworm treatment is available, but it must be closely supervised and can take many months before your dog is declared healthy. The treatment itself is not without risk as adult worms and larva killed by the treatment die inside the heart and blood vessels. Side effects are relatively common even during successful treatment, and dogs usually require hospitalization as a precaution. The majority of treated dogs recovers well.

The life cycle of the heartworm takes place almost entirely in the circulatory system of the infested animal. It is transmitted in an early larval stage by the mosquito. When a mosquito bites a dog with heartworm larvae in the bloodstream, they are harbored by the mosquito until it bites another victim. When a mosquito bites an animal, the mosquito initially expels some saliva that keeps the victim's blood from clotting. It is at this moment that a heartworm larva is also expelled to continue its own life cycle inside the animal.

Thick coats, as on this Chow Chow, do not protect against mosquito bites that could transmit heartworm disease.

Heartworm preventative pills are usually given once monthly on the same calendar day of every month during mosquito season. Because there is always even a remote risk of mosquitoes, more and more veterinarians are advising heartworm pills year-round. Many pet owners forget to go in for the heartworm test or become confused as to when to start and when to stop these pills. If you have stopped the heartworm pills for any reason, the blood test must be repeated. It can be simpler and safer for most pets and their owners to practice prevention throughout the year.

Cats do not get heartworm.

Cats can become infested with heartworm although this is not a very common problem. In fact, heartworm can even be transmitted to people by mosquito bites, although this is relatively rare.

Heartworm infestation can become a debilitating disease much sooner in cats than in dogs simply because the cat's heart is so much smaller and cannot harbor many adult worms. If your veterinarian suspects your cat might have heartworm disease, a blood sample will be recommended. If feline heartworm disease is confirmed, the treatment may be surgical or medical. Some cats actually require surgical removal of the worms from the heart. Other cats are left with adult worms inside the heart, but take medication that prevents additional larvae from growing. These cats survive for years despite their unwelcome guests.

The hair in a dog's ears should always be plucked.

In general, if something is not broken you shouldn't fix it. A dog with hair in the ear canals but no sign of otitis (ear infection or inflammation) should not have the ear hair plucked.

Sometimes, plucking the hair from a healthy ear irritates and inflames the canal and could even predispose your pet to infection. Some cats have abundant hair growth on the inner surface of their ears but the ear canal is usually quite hairless. Cats do not need to have their ears plucked.

If your dog develops an ear infection, you may be advised to keep the ear canal free of abundant hair growth. In the presence of infection, excessive hair can prevent a desirable flow of air into the canal to promote healing of many ear infections. Hair can harbor the infectious agents and delay response to treatment. Ears can be plucked by a professional pet groomer or by you after careful instruction by your veterinarian. If you prefer, your veterinarian will gladly do this at periodic intervals if your pet is prone to ear problems. Plucking the fine hair from the canal is usually simple and painless. Be careful not to pinch the skin with whatever instrument you use (a tweezer will usually do the job well). If you suspect your pet already has an infected ear, ask your veterinarian to examine your pet before you intervene at home.

Certain dog breeds require cropped ears or docked tails for health reasons.

The practice of surgically altering ear shape and tail length began in a time when dogs were used for working or hunting large wild game, or were legally "pitted" against each other in dog fighting arenas.

In this day and age, there is no value or health reason to justify cropping a dog's ears or docking the tail. These unnecessary and painful surgeries do not prevent medical problems and are rarely advocated by veterinarians. These procedures are still grudgingly performed (tail docking in pups between two and five days of age, ear cropping about three months of age), mainly to accommodate pet owners or breeders that aspire to traditional standards of breed appearance. As ear cropping and tail docking in many breeds loses popularity, these procedures may become obsolete some day.

Ear cropping and tail docking once were common in many breeds, but increasingly are considered unnecessary surgeries.

Dogs and cats eat lawn grass or other plants because they need to vomit.

Dogs and cats eat grass in your yard primarily because it is there. Young animals investigate the world around them by watching, listening, smelling, touching and tasting almost everything they encounter. Although our pets are classified as carnivores, or meat eaters, it may be natural to supplement their diets with plant materials. They can develop a preference for the taste or texture of an alternative food. Nibbling on plants is probably unrelated to any dietary deficiency and does not imply a need for more bulk or fiber content in their diet.

It is common to see dogs or cats occasionally eat some form of plant matter and regurgitate moments later. This suggests, but does not prove, that the individual intentionally sought out and swallowed the plant to induce vomiting. It is equally possible that an animal feeling nauseated or bloated might seek to ingest something it might not otherwise eat.

Eating plants may be normal in healthy pets. It is also seen in individuals with a variety of medical disorders, although this is not a specific marker for any one disease. Abdominal discomfort caused by the ingestion of spoiled food or the presence of internal parasites, for example, may be associated with a pet's interest in consuming unusual things. On the other hand, gastrointestinal distension could suppress a pet's appetite altogether. If your pet's attraction to lawn grass or other plants occurs daily and is associated with vomiting, diarrhea or weight loss, your veterinarian should be consulted without delay.

Sneezing is a sign of allergy.

Most of the time, excessive sneezing in cats and dogs is a sign of an irritation, inflammation or infection of the upper respiratory system (involving the nasal passages and sinuses).

Allergy in cats and dogs more frequently affects the skin or digestive tract. Individuals allergic to seasonal pollens, for example, are more likely to feel very itchy and scratch or chew at body surfaces. Food hypersensitivity occurs in pets and can be difficult to distinguish from a long list of ailments that cause itchiness and coat problems. It can also be present as diarrhea and other nonspecific signs of gastrointestinal upset. Allergy to medication can trigger skin, digestive and respiratory problems.

Respiratory signs that are so commonly related to allergy in people do occur in pets. Allergy-induced sneezing in dogs is associated with seasonal pollen although the skin symptoms usually predominate. Feline asthma is characterized by wheezing and moist coughing. It is thought to be an allergic phenomenon, but feline asthma may also be triggered by emotional stress such as reaction to separation. If your pet is sneezing, try to record how many times during a twelve or twenty-four hour period. Is there any nasal discharge and if so, is it cloudy or clear? Do the eyes seem affected, too? Is there any change in appetite or level of activity? If sneezing is complicated by any other signs of discomfort or disease or if the sneezing lasts for more than two days, consult your pet's doctor.

Colds are passed between people and pets.

The common cold (upper respiratory infection) in people is actually caused by many viruses. In general, the viral symptoms run their course and do not require anything more than optional symptomatic medication, unless complicated by a secondary bacterial invasion. Colds are usually of little concern unless the patient is less resistant to illness. Animals, as people, can get viral upper respiratory infections. Many of these can be prevented by making sure your pet is vaccinated every year.

Colds caused by viruses are not transmittable to other species, because upper respiratory viruses have generally evolved to affect specific species. Bacteria are somewhat less "choosy' about where they settle. Bacterial respiratory infections could be, in theory, passed between people and pets, but the chance of this actually occurring is minimal in healthy individuals.

Brown discharge from a pet's eyes is a sign of infection.

Brown stains on the inside corner of the eye is normal in dogs and cats. The reddish brown stain is caused by the presence of porphyrins, a substance normally found in their tears. When tears mix with any mucus flushed from the eye's surface and then dry, the crusty deposit can have a dark brown to black color. This, too, is completely normal. Unless the discharge seems excessive to you or is accompanied by signs of discomfort, ocular tearing and slight discharge are usually normal. You should be concerned, for example, if your pet is blinking or winking the eye, or if the white area of the eyeball looks pink or the inside of the eyelids look red. See if your pet is rubbing paws or face against furniture or carpet. In addition to these signs, if your pet avoids bright light, seems withdrawn and is less hungry than usual, see a veterinarian immediately. The eyes are delicate and can be affected by a long list of disorders that fortunately often respond to early treatment.

Mange in dogs is always contagious.

The skin infection referred to as "mange" is really at least three different conditions. All forms of mange are caused by microscopic insects called mites that inhabit the outer layers of the skin to cause a superficial or deep dermatitis. Mange can leave the pet with thick and crusty scabs, varying levels of slight to severe discomfort, and local or generalized hair loss.

Mange due to the *demodex mite* is not contagious to people or to other pets. It is suspected to have a congenital basis but is frequently activated by some other latent medical problem such as internal parasites or viral incubation. Mange due to the *sarcoptes mite* is transmittable to people although the condition is self-limiting and usually does not require treatment. Sarcoptic mange is also contagious between dogs.

Mange is diagnosed by scraping the surface layers of the skin with a scalpel blade. This allows the identification of mites under the microscope so that the appropriate treatment measures can begin. Demodex is easily revealed while sarcoptes is difficult to find. In some cases, skin biopsy is required to confirm a diagnosis by revealing mites in deeper layers of skin. Treatment can be complicated and prolonged, particularly when there is secondary bacterial infection. Any related medical disorder must also be identified and resolved. New medications for and treatments of the mange complex are available, and, administered with patience and love, the pet's recovery can be expected.

Ringworm in people is always caused by contact with cats.

Ringworm is not a worm at all. It is the name given to a common superficial skin infection caused by fungus in people and other animals. The lesion is typically round, dry in the center, and with a reddish boundary. Thick, grayish, crusty scabs may cover the surface and hairs adjacent to the infected patch usually fall out. The animal may or may not seem to itch or scratch at the area. There may be one sore or several, ranging in size from smaller than a dime to several inches across.

Ringworm is caused by several different fungi, any of which can be transmitted between people and pets. The infection is usually self-limiting, but requires treatment just to be sure it does not spread or worsen. The fungi that cause ringworm are often normal inhabitants of the skin. Some of these fungi are more frequently found on cats, others on dogs or people.

All animals, including cats, normally have fungi and many different types of microscopic organisms on the surface of their skin. Superficial fungal infections can occur in pet owners who are less resistant due to a preexisting illness, in very young or very old people, or individuals that have a microscopic scratch or just dry skin. Fungal infection in people may be transmitted from pets with active symptoms of ringworm. Some pets are blamed as the source of their owner's ringworm lesion even when they themselves have no skin eruptions. People are more likely to contract ringworm from contact with another person or by gardening without wearing protective gloves. An individual cat, dog or person with skin lesions should be treated to make sure the infection does not progress and is not transmitted. A cat that has no obvious signs of ringworm does not require medication unless the owner is repeatedly infected.

Cats cause AIDS (Acquired Immune Deficiency Syndrome).

The virus responsible for the disease complex known as AIDS is a very specific infectious agent that affects the human body and no other. It is a fragile virus in the environment and is transmitted by direct and immediate contact with a new potential victim. There is no proof that the HIV virus that causes AIDS is capable of infecting or being transmitted by cats or dogs. It is an exclusively human tragedy.

Leukemia is a type of cancer of the blood that exists in people and in many other animal species. Although the cause of leukemia in people is still unclear, feline leukemia is viral in origin. In cats, two types of leukemia are known to be caused by two different viruses, the feline leukemia virus (FeLV) and the feline immunotropic virus (FIV). Neither of these viruses is contagious to people.

Unfortunately, some people (including veterinarians) refer to the FIV infection in cats as the "Feline AIDS" virus. This misleading and inaccurate label causes many people to worry unnecessarily. The FIV virus should not be called anything else.

There are some striking similarities between FIV, affecting only cats, and HIV, affecting only people. First, the viruses belong to the same family. This does not make them identical or interchangeable. Both infectious agents can lay dormant for years before causing any physical symptoms. The disease processes that these viruses trigger are largely due to massive suppression of the individual's immune system, leaving them vulnerable to unusual infections. The similarity between these viruses is a basis for comparative research in human and veterinary medicine that may ultimately benefit both species.

Teeth and gum problems do not affect a pet's general health.

Dental and periodontal disorders are uncomfortable and painful. These can impact a pet's health by decreasing appetite overall. Infections anywhere in the mouth can directly affect the individual's health. Bacteria can enter the bloodstream from the mouth and spread rapidly to other internal organs such as the heart and kidneys.

Finally, malignant oral tumors can be extremely aggressive and can spread to other parts of the body, like liver and lung tissue, even before they cause discomfort or are detected. Healthy teeth and gums are essential to your pet's physical well-being.

Bad breath is a sign of dental problems.

Bad breath can be associated with gingivitis and dental abscess or tonsillitis and pharyngitis. It can also be a sign of metabolic diseases (diabetes mellitus, kidney and liver disorders), respiratory illness (upper respiratory infections, pneumonia), gastrointestinal diseases and immune-mediated disease. Almost any severe illness, like infections (viral, bacterial, fungal) or cancers, can affect mouth odor regardless of their actual location.

Diet is also important and can affect breath in many ways. A pet might consume decomposing debris in your yard, its own stools or the feces of other animals. Malnutrition and related vitamin or mineral deficiencies can also cause bad breath. Halitosis can also be the result of self-grooming, for example, if your pet has an anal sac infection or lower urinary tract disease.

If your pet has bad breath, it is important to find out why. Although a variety of products are marketed to mask mouth odor, these items are usually ineffective.

Veterinarians are not "real doctors."

Many people are surprised to learn that doctors of veterinary medicine undergo the same rigors of selection and training required of physicians to complete their formation. There are only twenty-seven schools of Veterinary Medicine in the United States and just four in all of Canada. Competition for the limited number of spaces at each school has raised the caliber of every veterinary candidate. Unfortunately, because of these limitations, many applicants are denied their dream of becoming a veterinarian.

The course itself is a grueling four years of graduate school and resembles medical school in many ways. Students of human medicine need study only one species, whereas veterinary students must become experts in many animal species! To keep things in perspective, it helps to remind ourselves that human medicine is, in essence, a subspecialty of veterinary medicine!

Veterinary medicine is not as advanced as human medicine.

Advances in human medicine are due to years of research in laboratories around the world. Every medication and every surgical technique is first applied to experimental animals. In theory, therefore, veterinary medicine is at the forefront of all medical knowledge. In practice, unfortunately, the potential of veterinary medicine is impeded.

Most pet owners, regardless of their desire, are simply unable to pay for the treatments that may be available to their sick or injured animals. The average veterinarian in a local clinic, regardless of professional training, often cannot afford to offer the expensive apparatus or diagnostic equipment available in most human hospitals. Referral to veterinary teaching hospitals

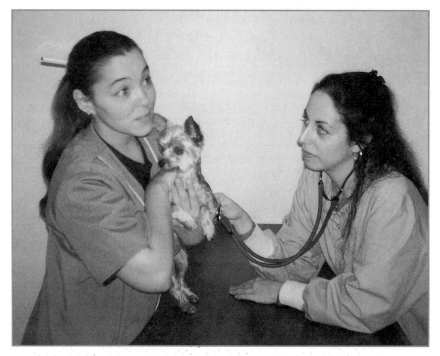

The author (right), assisted by technician Ruthie, examines a Yorkshire Terrier that had been aggressive until owners left the room.

and specialty practices where these options are offered may be financially or logistically unrealistic for most pet owners.

Many veterinarians complete an internship or residence, although this is generally not a requirement for private practice. More and more, graduating veterinarians pursue postgraduate specialty training. There are veterinary neurologists, pathologists, radiologists and orthopedic surgeons (with subspecialties in companion animal practice or large animal practice). Veterinary ophthalmologists, dermatologists and cardiologists offer advanced care, too. The field of veterinary behavior is parallel to human psychiatry and veterinary oncologists supervise chemotherapy to treat cancer in animals.

Some have observed, with humor based in wisdom, that physicians are "ordinary" doctors while veterinarians are "extraordinary" doctors!

Feed your pet a good quality food at regular mealtimes and keep fresh water available at all times.

13

Nutrition

Pets instinctively eat what they need to balance their diet.

"Nutritional wisdom" in pets does not exist. There is only sparse and inconclusive evidence in support of an animal's choice of food based on an instinctual knowledge of nutritional needs. The most that can be said is that animals tend to consume more if a food is lower in calories, and those with marginal metabolic deficits (in NaCl or NaHCO3) do tend to select saltier foods. Individual cats and dogs may have taste preferences for certain foods. Cats and dogs may refuse to eat one food because of a taste preference for another, or they may acquire an aversion to one food and refuse to eat it or anything that resembles it.

Most cats will prefer fish to meat despite the fact that a heavy fish diet has been linked to certain medical problems in cats. Dogs are particularly attracted to sweet tasting things, even though consumption of engine anti-freeze is lethal even in minute quantities.

Cats and dogs only eat what they need, so food can be available constantly.

Self-regulated feeding does not really exist. Although most cats maintain a fairly constant body weight during their lifetime, approximately 10% of cats will overfeed and become obese if given continuous access to food. In dogs, the tendency to eat may be more related to the time of the last meal, and they are even more prone to obesity.

Feeding is primarily intended to sustain life. However, it is also an activity unrelated to hunger. Pet owners frequently give food rewards to pets for performing a variety of behaviors. Begging for food, for example, soon becomes a gratifying activity even when the pet is not hungry. A pet can also independently discover that asking for food is a way to gain your attention. Unfortunately, the "trick" or other attention-seeking pattern can be something that is annoying, such as a dog's persistent barking.

Feeding is socially facilitated, that is, it may be triggered by the presence of other individuals. This is not unlike the favorite human activity of family dinners or going out to dinner. In pets, too, socially facilitated feeding occurs when familiar individuals are nearby.

Negative social factors, however, may also trigger overeating.

- *Psychogenic hyperphagia,* or excessive eating due to emotional causes, has been associated with the introduction of new pets or new people to a household.

- Cats and dogs may eat and drink more than usual in times of stress.

Conversely, a decrease in appetite may be the only obvious sign of physical illness in pets. Many parallels exist in eating disorders of both people and pets. For most pets, the best feeding method is to provide two premeasured meals at regular intervals. Remove any uneaten portion. This will make it easier to balance food intake with exercise in order to maintain a healthy body weight. This method of feeding also helps to regulate bowel habits to maintain house training in dogs and minimize finickiness in both cats and dogs. Weight maintenance is determined by the amount of exercise as well as the quantity and the quality of the diet.

Adult cats and dogs should only eat one daily meal.

If dogs and cats were permitted to feed freely they would probably consume about one dozen small meals throughout the day. Cats would feed at any time of day or night, whereas dogs generally tend to rest during the night.

To maintain a pet's ideal adult weight, however, it is probably unwise to indulge their natural rhythms. Free-feeding (unrestricted access to food) will undoubtedly cause obesity for most domestic pets. On the other hand, restriction to one daily meal may not be easily tolerated by some pets. It might be difficult to watch their owners prepare and consume several meals each day without anticipating some food of their own. Twice daily feeding gives them two special daily events to look forward to and also distributes calories to better coincide with periods of arousal.

Obesity is the single most important health problem among pet cats and dogs.

Pets that become anxious or competitive with each other at mealtime may be less stressed with multiple feedings. Remove any uneaten portion to regulate weight control. There is nothing really wrong with feeding a pet only once a day, however, two daily feedings are better in the long run for the average house pet.

More female cats become overweight compared to male cats.

There is no evidence that indicates a gender-based predisposition for obesity. No significant difference has been found between the number of overweight male and female cats. The quality and quantity of food, as well as activity level, are far more important.

Although there are fewer obese cats than dogs, fifty percent of house cats, compared to thirty percent of outdoor cats, become overweight. This does not advocate allowing pets to roam outdoors in order to control weight gain. The obvious implication is that house cats require closer supervising of food intake and compensation for outdoor activities.

The eating habits of pet owners do not affect the eating habits of their pets.

Pet owners absolutely do influence their pet's eating habits. There are very few of us who truly never ever give our pets scraps from the table or bring home a "doggie bag." The eager and pleading expression of a dog or cat anticipating a taste of our food is difficult to refuse.

This is one way that finicky eaters are created. Many dogs gradually learn to refuse dog food when they are fed home-cooked meals. Owners of small and Toy breeds, for instance, are particularly guilty of this. Another way in which pet owners affect their pet's diet is by imposing on them their own philosophy of eating.

Some pet owners follow vegetarian diets and impose these on their animals. Our pets are carnivores and require meat as the basis of their diet. It is also recognized that pet owners who are obese may have obese pets, especially if they are also elderly. To be fair, an obese person can be the most concerned about controlling a pet's weight and do a better job at it than pet owners of average weight. For ourselves and our pets, we must admit that food is not love.

Cats should be allowed continual access to food.

Many cat owners fill bowls for their cat's continual access to food. This practice predisposes many pets to gradual weight gain and, eventually, obesity. Cats confined indoors as house pets may nibble around the clock even when they are not hungry. Outdoor cats may be less prone to obesity but can scavenge or hunt for food outside and then return home to eat even more. If feeding is *ad lib,* a pet may overindulge when passing by the food bowl or as a recreational activity when there is nothing better to do. Cats can learn to ask for more food when they are not hungry and just want attention from their owners. Allowed continual access to food, most cats will eat about one dozen small meals day.

Studies of feeding in big cats, such as lions and tigers, indicate that their pattern is to gorge themselves on one *occasional* large meal. In contrast, the natural pattern of the domestic cat may be to feed on many small prey, such as mice throughout the day.

The number of meals each day may be more related to the effort required to obtain them. Adult cats should be fed two daily meals at approximately twelve-hour intervals. Feed a good quality food recommended by your veterinarian in keeping with the cat's physical condition and activity. The appropriate quantity is that which maintains an ideal weight. Package labels are only guidelines. The only reliable indicator of how much to feed your cat is your cat's appearance. An obese pet is not a healthy pet because obesity is a disease. Your pet will also risk developing several obesity-induced illnesses if excessive body fat remains over time.

Cats will not eat anything outdoors if they are well fed at home.

Cats that roam outside will eat whatever food they find. They may raid your neighbor's (or your own) trash or scavenge the remains of school yard lunches. They are free to express their individual inclination to hunt and consume their prey. Although cats that go outside have many more opportunities to burn off calories, this does not mean that they won't get fat. They may very well eat outdoors and then indulge in whatever feast awaits them at home. Most pets will not self-regulate their food intake. If your outdoor cat is gaining weight, you can either restrict its ability to go outside or you can control what you feed it indoors. At the very least, a gradual introduction of a low calorie commercial diet is advisable. Adjust the quantity of any food to your cat's appearance. Remember, overfeeding "diet" food will maintain or add to body fat.

Male cats should not eat dry cat food.

Urinary tract disease in cats is the subject of ongoing debate and much remains unclear.

Many things are suspected of contributing to the complex of Feline Urinary Tract Disease (FUTD), which was formerly labeled Feline Urological Syndrome (FUS). Predisposing factors include:

- Heredity;

- The amount of magnesium (Mg) and phosphorus (P) in the diet;

- Water intake;

- Urinary pH;

- Underlying viral infections;

- Subsequent bacterial infections;

- Poorly managed litter boxes;

- Inactivity and obesity.

Some pets that are fed dry cat food never develop FUTD while others that have never even tasted dry food suffer from repeated episodes of discomfort. Symptoms of FUTD are often related to the presence of bladder crystals but can also occur in the absence of crystals. Urinary crystals are frequently found in cats that never have symptoms of FUTD. The formation of crystallized mineral deposits in the urinary bladders of predisposed cats has been attributed to a variety of things. The "ash content," which refers to the balance of Mg and P in a food, may create a basis for Mg-based crystals. Since the discovery of the importance of the Mg to P ratio, many (but not all) commercial pet foods have been reformulated. Prescription cat food diets have been of tremendous help in treating cats predisposed to urinary crystals caused by diet.

Cats confined as house pets may be more prone to developing FUTD because they are at higher risk of becoming obese, too. Both males and females are equally at risk of developing FUTD, regardless of whether they are neutered, or what age they were neutered. The consequences are quite different, however, because FUTD in female cats rarely results in urinary blockage. In males, deposits of crystals or other debris, spasm and inflammation of the urethra can cause fatal blockage. Some male cats that eat dry food exclusively may never develop symptoms of FUTD, while some that eat only canned food require emergency care for urinary blockage. Not every male cat that is diagnosed with FUTD will become blocked.

Until risk factors are better recognized, cat owners might consider the following:

- Keep male and female cats lean by balancing food intake with exercise;

- Encourage them to stay active by inviting them to play;

- Change water bowls daily and even add water to food;

- Keep litter boxes clean so that cats will not hold their urine because of aversion to an unclean box;

- Avoid cat foods that are high in fish or seafood or as directed by your veterinarian.

Pets know when a food is not safe to eat.

Despite sensitive noses and acute vision, our pets can fail to detect a food that is spoiled or contaminated. Compared to most dogs, cats tend to be more selective about what they ingest. Nonetheless, food or water can contain hazardous substances, such as lead, that are tasteless.

Protect your cat and dog by acting preventatively whenever possible. Buy pet foods that are known for quality and quality control. Popular brands will have a higher turnover on store shelves. Walk your dog on a leash and look out for objects that dogs could find attractive along the way. Put poisonous automotive or household items safely away. Some substances that are tasty to pets, such as car antifreeze, are deadly. Petproof your home and yard much as you would childproof the area. Innocence must be protected.

Strictly vegetarian diets are healthy for dogs and cats.

Dogs and cats are carnivores. Their digestive systems are intended to digest meat. Proteins found in meat are essential for the development and maintenance of healthy nervous systems, muscle, heart, eyes, skin and coat. Pets can relish the taste of fish or cheese and other sources of protein, but the basic dietary requirement for red meat or poultry has no substitute for dogs and cats. Of course, a strict meat diet, however, is no better for a carnivore than a strict vegetarian diet.

In recent years, overindulgence in meat has been proven to be unhealthy for people. However, the amount of meat we eat should not influence what we feed our pets.

People are omnivores. That is, we function on a mixed diet incorporating a variety of foods. We can choose to include or exclude meat as a source of protein as long as we select other food sources to compensate. Simple dietary substitutions do not apply to carnivores. If you have serious personal objections to feeding your dog or cat a meat-based diet, please consult your veterinarian who can guide you toward a compromise of a kind with careful dietary supplements. As a loving pet owner, despite your personal philosophies about food, consider what is ultimately best for your pet's health.

All pet foods are alike—the best brand is the cheapest.

The pet food industry is one of the most lucrative in the world. In Canada, the quality of pet foods has been monitored for more than 20 years by the Canadian Veterinary Medical Association to ensure strict standards. In the United States, however, pet food manufacturers need only comply with nutritional guidelines set by the National Research Council.

Package labels must state that the food meets with the minimum amount of nutrients recommended by the NRC guidelines. No further proof of quality is currently required. Additional approval may be granted by passing tests of the American Association of Feed Control Officials, but these are optional. Pet foods must also be registered at state agencies. At first glance, this system might imply that the industry is regulated. Unfortunately, the rules fail to certify the quality of manufactured pet foods. American consumers must rely on the integrity and knowledge of pet food manufacturers to provide healthful diets.

Although a comparison of product labels can show identical percentages of protein or fat content, for example, all pet foods are not alike. At this time, pet food labels are only required to indicate a list of ingredients, a declaration of their minimum guaranteed content and a statement required by the NRC of "nutritional adequacy." Labels define contents as "not more than" or "not less than," but this does not reveal the exact amount of any given ingredient.

Ingredients are listed in decreasing order of content. The percentage of beef, however, does not tell you the quality of the meat used. For instance, thirty percent protein in one brand may be superior in quality to thirty percent in another brand. If the figures were calculated without the water content of the beef, the difference in nutritional analysis would show the actual amount of beef protein that contributes to actual nutrition. Even if brands seem to contain identical amounts of protein, the source, quality and method of preparation will determine which product is superior. Finally, don't forget that every individual digests food differently. Even if a food quality is indeed superior, your pet's individual digestive system might still be unable to digest or metabolize it. The best quality food may not be best in meeting your pet's nutritional needs. Pet owners should choose products endorsed by veterinarians and manufactured by well-recognized companies. Write to the pet food manufacturers and request a chemical analysis report of their product. While you are at it, write a letter to the NRC and ask for stricter regulations.

Beware of bargain brand pet foods because you may get what you paid for. Compare prices and note that some marketing strategies intentionally overprice pet foods of only moderate quality. The best testimonial to a pet food remains, of course, your pet's appearance and preference.

Pet food labels reliably determine your pet's daily food intake.

The best way to determine how much to feed your pet is to look at your pet. If your pet is gaining weight, you should increase daily exercise and cut back on its food intake. If it is losing weight, you should increase caloric intake. *Use the package label as a guideline only.* If there is any unusual weight gain or loss, you should certainly consult your veterinarian without delay.

Even if a pet is fed a high quality food, the digestion of the food is determined by the individual's metabolism. The pet's level of activity will, in turn, decide if calories are stored in fat or not. The balance between food intake and body weight is only partly a function of your dog's age, for example, or whether the animal is neutered or not. The rest is up to you.

Pet foods labeled as "natural" or "health food" are better quality.

The use of terms that declare a food product to be more "natural" than others is the subject of ongoing controversy even in foods marketed for human consumption. Health foods are not necessarily of superior quality. In some cases, they may even be inferior to foods that are not additionally proclaimed as natural or healthful.

The addition of preservatives to commercial pet foods does not mean they are unnatural. Common food preservatives include the antioxidant vitamins.

Food preservatives are extensively tested for safety and effectiveness. These are frequently important to prolong the shelf life of commercially prepared foods and prevent serious illness from the ingestion of spoiled food. Some foods contain additives that enhance the appearance of the product by giving it a particular color or consistency. These products are not of any advantage to the pet but are marketing strategies directed at pet owners.

The best food is the one that maintains good health and that your pet will eat. Items advertised as "natural" or "health food" as a marketing ploy are frequently overpriced and do not prove quality. The most popular pet food brands are usually among the best quality.

Compare package labels but understand their limitations. Protein can be of very different quality, but this will not be indicated by the figure on the label. If content information is absent, it might be wise not to purchase the product. If two products appear to have the same protein content and you are unsure which to select, choose the better known brand.

Dry food prevents dental problems.

(See also chapter 14 "Grooming," section entitled *Eating dry food takes care of a pet's basic dental care.*)
There is no evidence that eating dry food will prevent tooth or gum disease in either dogs or cats. Some pets eat exclusively canned or semimoist food and develop few dental problems. Others fed strictly dry kibble diets can have problems such as tartar buildup (dental "calculi") on tooth surfaces (and under the gum line), staining and decay of the enamel and gingivitis. Certain breeds may be predisposed to dental problems regardless of their diet. Miniature Poodles and Persian cats, for example, are more prone to

dental calculi compared to other breeds. The only food item that has been shown to slightly reduce tartar in dogs is the rawhide bone, especially in the shape of a long and flattened chip.

Recently, a new line of prescription dog food claims it has a revised consistency that reduces the accumulation of surface tartar. Although the company is well respected and has some laboratory evidence to support the advertised claim, it remains to be seen how well this holds in the real world.

In pets as in people, mouth disease can lead to halitosis, discomfort and even pain, as well as serious general health problems. Variations in the pH of saliva and the function of salivary glands have been related to dental health. Inflammatory gum disease can severely affect pets. These may be inherited or secondary to many underlying medical disorders, which may include the feline leukemia viruses, kidney or liver disease, diabetes mellitus or infections due to foreign bodies embedded in the oral mucosa.

The best way to prevent dental disease is to have your pet examined regularly by a qualified veterinarian. A thorough investigation of all tooth surfaces should be part of every cat's and dog's annual checkup. Some forms of tartar can be manually removed by your veterinarian, but when this is not possible or if pets do not tolerate manipulation of their mouths, anesthesia may be required. Brushing your pet's teeth has been shown to minimize tartar and promote healthier gums. Pet toothbrushes come in a variety of forms (the simplest one is soft and fits right over your finger), and flavored toothpastes may increase your pet's tolerance of brushing.

Gradually train your pet to tolerate the sensation of a gentle finger massage along the gum line and outer tooth surface. Next, praise when your pet will tolerate a soft moistened cloth or pantyhose wrapped around your finger to do the massage. Graduate to a pet toothbrush or continue to use the cloth and add a paste made from a mixture of baking soda and water or use a pet toothpaste.

Dogs must chew on bones as part of their diet.

Although bones contain essential minerals required by dogs and cats, these elements are provided in the safer form of bone meal in pet food. Pets derive great satisfaction from chewing on bones. Chewing on hard objects strengthens muscles in the head and neck and provides a challenging recreational activity. Chewing can help to minimize the accumulation of dental tartar depending on the object's shape and the pet's individual technique, of course.

Unfortunately, this instinct can also predispose them to very real danger. Chicken and other poultry bones are porous, as are the bones of all birds. These splinter easily into daggerlike objects that can lacerate your pet's digestive tract. Therefore, they should not be offered. Similarly, smaller steak or chop bones should be restricted, but even the bigger marrow bones and massive knuckle bones can harm your dog. A dog's upper or lower jaw can get stuck through the contours of ring-shaped bones. Bone fragments can become lodged in the mouth or throat and cause respiratory distress. If swallowed, intestinal obstruction may result. Even small bone fragments can cause constipation and discomfort. For these reasons, it is probably best not to give bones to your pet at all, or alternatively, to provide them for limited periods of time and only under supervision. Pets that become aggressively possessive with bones should be denied them altogether for everyone's safety.

Chewing bones is good for your pet.

While it is true that the mineral content of bones is important for your cat or dog's balanced diet, these minerals are usually contained as processed bone meal in pet foods. Mineral supplements for your pet may be recommended in specific medical condition, but this is unlikely to be necessary for a healthy pet. All bones can cause intestinal obstruction, or fragment and lacerate the bowel. These accidents may cause panic and pain for the pet innocently enjoying a treat and often require surgical intervention.

Of course, puppies and dogs need to chew on something. Offer them appropriate chewable objects made of rawhide or tough synthetics such as nylon. Provide them with a wide variety of chew toys of all shapes and sizes

Joey, a Golden Retriever, enjoying a rawhide bone.

to keep them occupied for just several minutes at a time or several hours. Remove chew toys if your pet has a tendency to swallow them whole when only smaller pieces remain. If your dog becomes aggressive when chewing on favorite items, you have several choices: eliminate the chew toys altogether; offer them only when your pet is certain to be undisturbed; or teach your pet to relinquish these and other objects on command. Finally, kitchen waste containing potentially dangerous but tempting bones must be secured and made unavailable to your scavenging pet.

Excessive water intake should not be permitted.

Many medical disorders can trigger an increase in thirst. Water restriction in a pet that drinks excessively could have severe consequences if, for example, it is an undiagnosed diabetic. The magnitude of thirst also varies with temperature, activity and the quality of foods. Food that is high in sugar or salt or that contains little moisture (dry pet food), for example, will increase drinking. Never restrict your pet's access to clean and fresh water unless directed to do so by a veterinarian.

Excessive water intake has been related to social stress in dogs and cats. Your pet may be anxious when a new pet is introduced, for instance, but "psychogenic drinking" usually subsides without intervention. Sometimes, increased drinking of either medical or psychological origin is also associated with more frequent urination or urinating in undesirable locations. Even if your pet is house soiling, water must not be restricted. If you suspect your pet's thirst has increased suddenly or progressively over time, consult your veterinarian.

Cats must drink milk because they do not drink water.

Cats are mammals. All female mammals nurse their young with milk produced from special "mammary" glands. The mother cat, or queen, nurses her kittens until they are old enough to eat solid food. Cats, therefore, do drink milk when they are newborn and as young kittens. *This does not mean that they do not drink water.*

In general, cats drink about twice as much water by volume as they eat in food. Cats that eat canned food drink less water compared to cats that are fed dry food–based diets. This is because the water content of canned foods is already very high. In cats with urinary tract diseases, however, your veterinarian may advise you to add water to canned or even dry food. After weaning, cats of any age should be encouraged to drink as much water as possible. Change drinking water daily to keep it fresh and be certain it is easily accessible.

Cats require milk in their diet.

The idyllic image of barn cats drinking milk from a milking pail lingers in our urban minds. However, the romance of country living distorts the true hardships of their farm existence. In rural fields or in urban alleys, stray cats must take their nourishment wherever possible. They are unlikely to enjoy regular meals of a quality commercial cat food and are not guaranteed access to clean and cool water.

Cats, like all mammals, require milk as newborns and through the early period of growth. Once kittens begin to ingest solid food, however, they can be weaned from a milk-based diet. Cats do not require milk in their diet if they eat quality cat food and drink water. They may continue to drink milk, however, if it is available to them. A cat that has not had access to milk or milk-based foods for a while may have diarrhea when these are first reintroduced. This is because the enzyme required to digest milk is produced in adult animals only if milk and related food items are a steady part of their diet.

Drinking milk causes worms.

The only way that drinking milk will result in intestinal parasites is if it has been contaminated with fecal material containing parasite eggs. If a cat or dog has not had a drink of milk in a long time, drinking milk may cause diarrhea because the special dietary enzyme called lactase required to digest milk ceases to be produced. The notion that drinking milk results in "worms" may come from the fact that they can both be associated with gastrointestinal upset. Until the body can respond to a renewed demand for lactase, diarrhea can result from incompletely digested milk products. If milk consumption continues, lactase production is usually reactivated and stools return to normal unless the individual is lactose-intolerant or unless it really does have "worms"!

A pet's diet must be varied to prevent boredom.

Our pets are fortunate in that they need not be concerned with balancing their own diets or with the psychosocial pressures of food as we are. They do not need variety in their diet if they are fed quality commercial foods that meet their individual nutritional requirements.

The only reason to introduce variety is to prepare your pet to more readily accept a diet change. This might become necessary, if your pet food brand is discontinued. Also, if you travel away from home, your brand could be less available. Your pet's health status might require a change to a prescription diet or restricted diet of some kind. There is no reason to change pet food brands as long as the food provides a quality diet that: (1) the pet consumes eagerly; (2) maintains the pet's physical appearance; and (3) meets the individual's nutritional needs.

It is not necessary to vary your pet's diet, but if you do, it is best to blend the new and old diets together for several days or more before gradually eliminating the old food from the mix.

Table scraps are important for a balanced diet.

To provide a balanced diet, all you need is to feed a quality pet food your pet enjoys. Some pet owners prefer to prepare a homemade diet instead of a commercial pet food. If this is your choice, consult with your veterinarian for nutritional guidelines and tips. Table scraps are unnecessary for nutrition unless they become part of a homemade diet.

Feeding table scraps is usually most important for a pet's owner. We are naturally inclined to share food with those we love. Overindulging a pet with table scraps can lead to finicky appetites and even in a refusal of pet food altogether.

The occasional table scrap can be included with a pet's portion, but feeding pets directly from the dinner plate is never a good idea. This encourages undesirable behavior such as whining and barking, which can even progress to jumping in your lap or onto the table! If giving your pets an occasional taste of "people food" makes you happy, mix a small taste of a bland treat (spicy or unusual foods can upset a pet's digestive system) with their regular food in their food bowl during the usual mealtime.

Adding brewer's yeast or garlic to your pet's diet will prevent fleas.

There is no scientific evidence to justify why garlic or brewer's yeast would ever be effective in preventing fleas. Some people swear this works because their pet does not seem to attract fleas. Any therapeutic effect attributed to brewer's yeast is entirely coincidental as the same pet owners will discover when their pet eventually does get bitten by fleas. Perhaps this common misbelief is related to the notion that garlic keeps vampires away.

Raw eggs ensure a healthy and shiny coat.

Pets require a balanced diet appropriate for their species, breed, reproductive status (e.g., growing puppies or kittens and pregnancy demand higher protein content), activity, age and general state of health.

Adding a raw egg to a pet's food is an unnecessary addition of protein and cholesterol. A healthy pet fed a commercially prepared or veterinarian supervised home-cooked diet satisfying the individual pet's needs does not ordinarily require any dietary supplement. In fact, raw egg contains an enzyme that destroys the vitamin biotin. Feeding raw eggs (particularly egg white) to your pet can cause physical disease due to biotin deficiency. This is easily avoided by not adding raw egg to food or, if egg is part of a home-cooked diet, by supplementing cooked egg to the diet.

A pet's coat and skin are often a reflection of general health. Almost any medical problem will affect the external appearance of the individual. Symptoms such as vomiting, diarrhea, fever and overproduction of urine can result in dehydration that contributes to the coat's dull appearance. Obesity, for example, can affect liver function, resulting in a greasy coat. Allergy, parasites and excessive bathing can lead to a dry coat with flaky and itchy skin.

Everyone's skin and hair are affected by overheated homes during the winter. In some cases, coat quality is poor even when a pet is being fed a premium food. Some individuals digest certain foods better than others and would benefit from a gradual change to another quality food. Instead of unnecessary and expensive diet supplements, discuss any of your concerns about your pet's health with your veterinarian.

Raw fish is healthy for cats.

Uncooked fish contains an enzyme that destroys the vitamin thiamine. A cat that consumes large amounts of raw fish on a regular basis will suffer the effects of thiamine deficiency. Clinical symptoms include a hunched appearance, a poor coat quality and a long list of neurological signs. If untreated, severe thiamine deficient states are fatal. Fortunately, this is easily avoided by not feeding your cat raw fish. The fish will be grateful, too!

An exclusive diet of cooked or raw meat (or liver) is good for pets.

A diet that consists only of meat is inadequate for cats and dogs. As carnivores, pets definitely need proteins derived from meat, but other ingredients are essential to meet their dietary requirements. Cats actually require a higher percentage of meat protein in their diets compared to dogs.

A well-balanced pet food should also contain fats, from which cats derive as much as sixty percent of their energy. Vitamins and minerals are also necessary. The addition of carbohydrates to pet foods may not be essential, but this is a less expensive way to supply some of their nutritional needs.

Pets that are fed only meat risk serious nutritional deficiencies that can result in any number of complicated illnesses. As a consequence of prolonged meat-only diets, the body activates mechanisms which attempt to compensate for the individual's nutritional imbalances. The disease called nutritional hyperparathyroidism is triggered by imbalance of essential minerals. This results in a dramatic decrease in bone density. The skeleton wastes away as the body tries to supply vital organs, such as the heart, with calcium. The bones become so thin and brittle that fractures spontaneously occur. A balanced diet is key to the prevention and treatment of nutritionally induced disease.

A home-cooked meal is better than commercially prepared diets.

Not all pet foods are adequate and not all home-cooked meals for pets are inadequate. Even with the best intentions, the average pet owner would probably soon discover that preparing a satisfactory homemade diet is too expensive and time consuming. The decision to provide your own pet food requires a long-term commitment because your pet may resist diet change.

For those who are willing to try, it is essential to consult your veterinarian for feeding tips. To determine individual dietary requirements, you will need to consider your pet's age, athletic fitness, type of daily activities, general health and preexisting medical conditions. It is very challenging to try to ensure all the vitamin, mineral, protein, carbohydrate, fiber, fat and water ingredients in a proper proportion that will also be palatable. Your best efforts may be limited by your pet's individual preference for taste and consistency as well as how digestible the food is.

Consider your reasons for wanting to feed a home-cooked diet. Refusal of pet foods can occur because of abrupt diet changes, but is also part of normal fluctuations in appetite. If your pet is increasingly or persistently finicky, discuss your concerns with a veterinarian in case a medical explanation is uncovered. If you object to the quality or to a particular ingredient in pet foods, there may be other options of which you are unaware. If your pet is ill and refuses prepared pet foods or requires a specially prepared meal, a home-cooked diet may be specifically recommended.

A teaspoon of cooking oil ensures a shiny coat.

The only thing that cooking oil will supplement is your pet's intake of fats. Extra fat-derived calories are not useful for coat or skin quality. Fats are essential to a pet's diet (cats require a higher percentage of fat in their food compared to dogs), but there are many types of fat and only specific ones will do. Natural oils are secreted by glands in the skin and protect hair and skin, composed of protein, from drying. Cooking oil is not a source of nutrition and is unnecessary.

Concerned about your pet's appearance? Speak with a professional. If your pet is found to be healthy and has a poor coat despite adequate nutrition, prepared supplements of the specific essential fatty acids might well benefit your pet. A healthy pet does not require any additional fat.

Cooking oil will relieve constipation.

Constipation is the delayed and sometimes uncomfortable elimination of hardened and dry feces. If your pet appears constipated but is otherwise eating and behaving normally, the first thing you should do is to increase its exercise and water intake. Should this fail or if you have any other suspicions, consult your veterinarian before doing anything further. Never treat your pet with any over-the-counter laxatives, enemas, prescription drugs or home remedies without professional instruction to do so.

Cooking oil has no laxative properties unless consumed in quantity. Excessive consumption of fats can cause signs of indigestion from gas (flatulence) to diarrhea in pets with otherwise regular bowel function. A large amount of cooking oil can certainly relieve constipation but it may do so drastically. It may also not work at all except to increase your pet's caloric intake.

Shedding hair can be prevented by medication or nutritional supplements.

Products that promise to stop shedding are misleading. Special sprays, shampoos and nutritional supplements are advertised but none are recommended. There is no substance currently known to stop the normal and desirable turnover of a pet's coat. Applied topically, such a product could even be harmful if it irritates the skin or triggers an allergic reaction. In oral form, there is no regulation of ingredients yielding unfounded promises.

Normal hair growth and replacement occurs cyclically. In most dog and cat breeds, hair grows and normally falls out, or sheds, about every 180 days.

Normal hair loss does not occur all at once so that the skin remains uniformly covered. Seasonal peaks of shedding typically occur in the spring and fall. Warm weather triggers thinning when hair is no longer required to conserve body heat. In the fall, hair follicles again respond to seasonal hormonal changes that stimulate more rapid turnover and growth in preparation for the winter.

The coat of pets that spend time outdoors often becomes more dense compared to those that remain indoors. In nordic dog breeds like the Alaskan Malamute, for example, the hair grows in two layers. The longer outer coat overlays a short, dense layer of hair that is cottony in quality. This "undercoat" provides additional insulation in severe winter cold. In the spring, the dense undercoat frequently seems to be shed in clumps.

Some pets shed very little and others hardly shed at all. Airedales and Poodles, for example, show little hair loss but still require periodic professional grooming to rejuvenate the coat. The Chinese Crested hairless variety is practically bare and has virtually nothing to shed. The Rex cat was derived from selective breeding of a particular mutation that limited the coat to only a fine undergrowth that requires little grooming.

Shedding hair can be prevented by special sprays or frequent bathing.

(See this chapter, section entitled *Shedding in dogs and cats can be prevented by medication or nutritional supplements.*)

Your pet cannot develop an intolerance or allergy to food it has been eating for a long time.

Food allergy can occur at first contact with a food. In pets, however, food allergy typically emerges during prolonged exposure to the allergenic diet. Symptoms of suspected food allergy may not become obvious until they progress to the extreme. Signs of dietary intolerance are easily confused with many other types of allergies and skin conditions.

Food allergy in pets can also resemble gastrointestinal upset and appear as intermittent vomiting, pasty stools or severe diarrhea. Because many diseases can share the same or overlapping set of symptoms, your veterinarian will require your patience and cooperation until a diagnosis of food allergy is confirmed. A long list of medical possibilities must be considered before concluding with certainty that your pet is indeed allergic to a food.

Food allergy is not as common in cats as it is in dogs. In both species, the food category that causes an allergic response is usually a protein. Pets can be allergic or intolerant to fish, chicken or beef, but they can also have a problem reaction to wheat, for example. One of the things you may be advised to do is introduce a new food to replace your pet's current diet. Commercial diets that contain lamb as the sole source of meat protein are frequently substituted in pets suspected of dietary intolerance. This is because most pets have never eaten a lamb-based diet until it is recommended and therefore are unlikely to have developed any adverse response to it. If symptoms of allergy stop with a diet change, a diagnosis of food allergy is likely. It is important to note that some pets can develop hypersensitivity to lamb. No food can truly be called "hypoallergenic" because allergy is an individual phenomenon.

Dogs eat stool because they have a nutritional deficit or abnormal temperament.

(See chapter 9, "Elimination Problems," section entitled *Coprophagia is corrected by treating stools with a chemical or distasteful substance or by changing the dog's diet.*)

Pets eat lawn grass and other plants because they have a nutritional deficit.

Dogs and cats are carnivores. This means that they evolved to eat a primarily meat-based diet. It is likely, however, that their wild ancestors occasionally

Occasionally, grass may be a normal part of a dog's diet.

supplemented their diets with vegetation. Dietary flexibility can be an important adaptation in times when usual food sources are scarce, for example, in times of drought. There is no evidence that eating plants is associated with any dietary imbalance or search for fiber.

Pets are often initially attracted to household plants because of their natural instinct to evaluate new objects. This is typical of young kittens and puppies, for example, that base their initial exploration of the world on oral investigation. They may then develop a preference because of taste or consistency for a specific plant that could well become part of their basic diet. If your pet's interest in your plants is simply a playful nuisance, place these plants out of reach or in a room that is off-limits. In some cases, it might be best to tailor your choice in houseplants to ones that do not attract your pet.

This behavior is usually not a health risk but should be discouraged if the plant is toxic to pets. Lawn grass is not poisonous. However, if it has been treated with insecticides, it could become a menace to your pet. Some household plants and some shrubbery and trees common in home landscaping contain substances that are toxic to pets. Your veterinarian should be able to inform you if your plants pose a problem.

Salivating (drooling) means your pet is hungry.

Saliva is produced by specialized glands at several locations in the head and neck that empty into the mouth. Salivation is triggered by a variety of senses and situations. The odor, sight and taste of food are high on the list of stimuli that induce salivation in preparation for feeding. This is because one of the functions of saliva is to begin the process of digestion even before swallowing occurs.

Salivation can also occur in response to things that are completely unrelated to eating. During periods of acute stress or fear, individuals can experience a sensation of dryness in the mouth, but occasionally the opposite can happen. Some dogs and cats tend to hypersalivate when they are fearful or anxious. In individuals who are stoic and show little external signs of anxiety or fear, sudden and abundant salivation may be the only clue to their internal stress.

Salivation can precede regurgitation or vomiting and is a nonspecific sign of nausea associated with gastrointestinal upset. In dogs and cats that are unused to car travel, for example, salivation may be emotionally induced by a fear of travel or physically induced by travel sickness or both.

Some dog breeds such as the St. Bernard and Basset Hound are known for drooling. They appear to salivate more than others, but the effect is purely mechanical. In dogs with heavy facial folds and drooping flews or lips, drooling is due to the accumulation of saliva into pockets formed by lip folds and the inability to collect it efficiently for swallowing. Drooling may be more pronounced in some individuals compared to others of the same breed. The constant moisture of the skin can predispose the dog to superficial skin infections and considerable discomfort. Daily hygiene may be all that is needed to control the problem, but antibiotics and plastic surgery to minimize skin folds can be required. Many breeds, characterized by loose skin around the lips and cheeks, are as charming as they are sloppy. Nobody's perfect.

Neutering your pet will make it fat and lazy.

(See chapter 10, "Sex-Related Problems," section entitled *Neutering will make a pet fat and lazy.*)

The Afghan Hound is an elegant, athletic and highly intelligent dog that requires daily grooming for its long and silky coat.

14

Grooming

Healthy pets have no body odor.

Our society and culture bombard us with messages that body odor is unnatural, shameful and unattractive. We cover our scents with perfumed soap, cologne, deodorant, mouthwash and aftershave lotion. We then impose the same impossible standards on our pets.

A clean and healthy animal has a normal individual scent. We can deny that we are animals, but we should not project this onto our pets! Frequent bathing is recommended for people, but not necessarily for other species. Bathing a pet in an effort to remove the natural odor also results in eliminating the superficial oils that protect skin surface. The luster of a healthy coat may disappear as the pet becomes prone to rashes and infections.

Pet owners usually become accustomed to their pet's individual body odors. This familiarity can be useful in detecting changes in health status. Ear infections, a common problem in many pet dogs and cats, may first be noticed by a change in odors near the head. Still, some odors produced by healthy pets are objectionable to their owners. If you object to your dog's odor, occasional application of "dry" shampoos marketed for human use might be helpful. You could try sprinkling a bit of baking soda on the dog's back and *brush it out*. Dryer sheets, intended to counteract static electricity in laundry, can be wiped over the pet's coat and may absorb some of the odors. Most people no longer notice odors that surround them continually. If you give it less importance, you will become accustomed to your pet's odor.

Dogs and cats, especially white ones, must be bathed at least weekly.

Pets with white coats tend to show more dust and debris than others with darker colored hair. White cats, like cats of any color, usually do a good job of keeping themselves well groomed. Dogs, on the other hand, are not as meticulous and frequently require human intervention. Regardless of coat color, however, bathing a pet once a week is excessive. It is not so much the frequent contact with water but the use of shampoos that can cause problems. Dogs bred for regular contact with water, such as Retrievers or Poodles, have natural oils that protect their skin from becoming overly dry. When shampoos are used, the surface oils that maintain healthy skin and hair are washed away, leaving dogs of any breed prone to skin infections. The coat can appear dull; skin can become dry and itchy. Shampoos should be used sparingly and no more than every two months, unless a therapeutic bath prescribed by a veterinarian is advised for a specific medical condition. The same is true of cats. Unless there is a particular reason to bathe a cat, for instance in the case of a flea bath or as directed by your allergist, cats rarely need to be bathed at all.

Trimming the hair away from a dog's eyes will cause blindness.

Many dog breeds have shaggy hair around their eyes. The Old English Sheepdog, Lhasa Apso, Bearded Collie, Soft Coated Wheaten Terrier, among many others, have soft hair around the face that can grow long enough to cover their eyes from view. Breed Standards and popular grooming styles for these breeds dictate leaving the eyes at least somewhat covered by facial hair.

There is no medical reason to intentionally cover a pet's eyes, in fact, the opposite may be true. Hairs can irritate the eye surface, causing corneal injury and infection. Hair covering the eyes can delay detection of ocular disease requiring immediate veterinary care. Hair is a distinct disadvantage if it obstructs a dog's visual field. Pets with hair-covered eyes possibly see less well and might misinterpret an attempt to pet them as a menacing gesture.

Trimming the hair around the eyes will allow the pet to clearly see the world as well as enable you to monitor your dog's health. You will also be able to recognize facial expressions more easily and detect a dog's focus of attention, fear or even impending aggression.

Cats do not need to be brushed.

Although healthy cats perform most of their own grooming, their owners should help them if possible. Cats normally clean themselves, removing dead hair or debris with their rough tongues. Hair that is removed is usually swallowed. In some cases, related in part to the amount of hair that is ingested, accumulated hair forms into compacted plugs that block the intestines. These "hairballs" can cause symptoms of nausea and vomiting and, occasionally, can progress to intestinal blockage that requires surgery.

Longer-coated cats have more of a challenge to keep their coats free of knotted and twisted hairs that quickly form thick and uncomfortable mats. They may also be more prone to hairballs. Cats (and dogs) shed all year long as new hair growth continually replaces older growth. In peak shedding periods of spring and autumn, gastrointestinal problems from hairballs are common.

Combing or brushing will help to keep your cat's coat free of matted hair and decrease the incidence of hairball-related problems. It is also a nice way for your cat to share quality time with you. Most cats enjoy being brushed, especially if they are rewarded and if brushing is begun at a young age. If your cat does not enjoy being brushed, try a simple round-toothed plastic comb intended for human use. Give your cat a small food treat or briefly play with the cat while you pass the brush or comb for only a few strokes. By associating with a positive experience, your cat will eventually learn that grooming is a positive experience and will tolerate it for gradually longer periods. If you know that your cat is intolerant to having certain areas combed, avoid them. If your cat is uncomfortable with belly brushing, don't brush there. Do the best you can within the limits of your cat's preferences. Both of you will find it more enjoyable that way!

Only long-haired cats suffer from hairballs.

The significant proportion of an average cat's waking moments are spent meticulously grooming the coat. During grooming, the rough texture of the feline tongue collects dead hairs to keep the coat and skin healthy.

Small amounts of hair are commonly ingested and pass through the alimentary canal to be eliminated in stool. Ingested hair can accumulate and cause problems, for example, during peak shedding periods in spring and fall or during emotionally triggered excessive grooming. Hair can also accumulate if intestinal transit is slower, as it may well be in obese cats or house cats that have limited exercise. If enough hair is swallowed, they can plug the digestive tract. The hair becomes compressed into the shape of the intestinal passage and, if regurgitated as it often is, resembles feces from a distance. The hairball "cough" is not a cough at all but a kind of gagging due to gastrointestinal irritation.

Hairballs can become lodged in the alimentary canal and cause obstruction. Intestinal blockages require immediate veterinary attention and, sometimes, surgical intervention. Because it takes fewer long hairs (compared to short hairs) to form an obstructing hairball, cats with longer coats may have

The cat's tongue is wonderfully adapted for grooming and feeding.

more difficulties. Cats with shorter coats groom themselves in exactly the same way as longhairs. Coat length is less important than how fastidiously the cat grooms, how much hair is swallowed at a time and how well intestinal transit functions to eliminate a bulkier load. Oral lubricating pastes, known as hairball "laxatives," promote easier passage of hair and are frequently necessary even when a healthy cat is regularly groomed.

A cat's tongue is like sandpaper.

The cat's tongue is a wondrous tool with many functions. It is a relatively short and muscular organ with a complex surface structure. The cat's tongue is adapted for grooming, feeding, tasting and even showing affection.

To say it is like sandpaper is like saying your birthday is just another day or that the Super Bowl is just another game. If you have the chance to examine it closely, you will see that it is smooth at the tip and around the border. Most of the remainder of the surface is covered with dozens of tiny hooklike protrusions. These little "spines" help to remove hair or debris during grooming. During feeding, the raspy tongue is particularly adapted to consuming larger prey and helps to strip flesh away from bone. Taste buds are also distributed across the tongue's surface. Taste, which relies mostly on the sense of smell, is so important in cats that they refuse to eat when they cannot smell their food. This might occur when sinus and nasal passages are blocked for example, due to an upper respiratory infection.

Cats groom each other in friendly and relaxed social interaction, promoting stable group dynamics. When a person is licked, this redirected grooming behavior is probably a sign of affection, but it could also be to sample the flavor of something on our skin (especially if we just ate a tuna fish sandwich). Regardless of why a cat may lick us with their delightfully rough tongues, it is an undisputed honor that a cat feels comfortable enough to choose us for this brief and intimate gesture.

Short-haired cats and short-coated dogs do not need to be brushed or combed.

Pets with shorter body hair may not require the same effort required to maintain the coat of longer-haired or shaggier breeds. This does not mean that shorter hair does not require care. Brushing or combing your pet helps prevent matting and tangling. Short-haired cats should be groomed at least twice each week to help prevent hairballs that result from the ingestion of hair during normal self-grooming. Short-coated dogs need grooming as indicated for the breed.

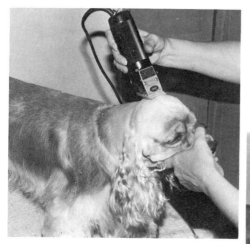

Professional dog groomers help you maintain your pet's healthy skin and coat, but bathing must not be excessive.

Brushing or combing also provides the owner with regular opportunity to inspect a pet for external parasites (like fleas or ticks), injury or growths on the body surface. Finally, when a pet is taught at a young age to enjoy being groomed, it gives both pet and owner a way to spend additional quality time together, strengthening their bond.

Fleas always make a dog or cat scratch.

The flea is a very small light to dark brown insect that feeds on the blood of cats and dogs. Visible to the naked eye, it remains on the animal's skin only as long as it takes to have a blood meal or lay eggs. Although much of this insect's life cycle revolves around your pet, most of its time is spent away from the host. If your pet has a flea, there may be only slight discomfort.

Fleas irritate the cat's or dog's skin by bites during the fleas' feeding. The sensation of fleas scurrying across skin can be increasingly uncomfortable. Some pets are allergic to flea bites, much as we are allergic to mosquito bites, except their reactions can be quite severe. These individuals can become extremely itchy and may lose hair over the fleas' target areas such as at the lower back. In sensitive animals, these reactions can occur even when there is only one flea. Like allergies in people, a flea allergy often requires the prescription of antihistamines or anti-inflammatory medications.

Not all pets scratch when they have fleas. Some cats and dogs can have dozens of fleas and show little or no reaction, or individuals can be severely infested with fleas and show no discomfort whatsoever. Regardless of a pet's tolerance to flea infestation, it is important to groom your pet regularly and to work toward consistent prevention.

Flea collars provide complete protection.

Flea collars are synthetic strips that have been saturated with insecticides active against fleas. They work by evaporation of insecticide that is randomly distributed over the animal's body surface.

Although flea collars sound good in theory, they are a frequent disappointment. Unfortunately, collars alone do not adequately protect cats or dogs from flea infestation. Under laboratory conditions, flea collars may perform well. In the real world of fluctuating conditions of wind and humidity, however, it is risky to rely solely on a flea collar to protect your pet. Flea collars combined with flea sprays, foams or powders provide more complete coverage that should be reapplied at regular intervals. Given a choice between collars or other flea control products, choose those which are applied to the entire body surface.

New "flea pills" completely replace traditional treatments.

A new generation of flea control products has recently flooded the pet product market. These compounds eventually render the exposed fleas infertile and affect their larva and eggs. An undeniable advantage of this product is that it poses no identifiable toxicity to people. Initially this makes it more attractive than insecticides currently marketed.

Administered orally once a month, the new compounds are deposited in the animal's skin and are ingested by the flea when it bites. This means that your pet (and you) will continue to be bitten and will still be exposed to the discomfort of fleas. Your pet will not be protected against other diseases carried by fleas, such as tapeworm and the microscopic blood parasite called "ehrlichia." New populations of fleas will not be prevented from invading your pet or home. These compounds do not kill fleas at any point in their life cycle. Fleas can live up to a year or more if environmental conditions are comfortable for them, regardless of whether they reproduce successfully. This compound does not prevent any fleas from inhabiting your pet or home and should not be considered a replacement for more traditional, albeit less environmentally friendly, products. People and pets that have flea allergies will find no relief if nothing other than "flea pills" is done to eliminate existing populations of fleas.

Another marketed form which incorporates these chemicals complements other more traditional items such as insecticide collars. This attempts to eliminate and to prevent *ongoing* flea infestation from harassing your pet. As discussed above, collars are still less effective than products that are more reliably applied directly to the animal's entire body.

Hopefully, sprays will soon be available that combine the flea "hormones" with insecticides for an overall less toxic product. In the meantime, flea control insecticides are nowhere near to being obsolete and their judicious application is still recommended.

All ticks carry Lyme disease.

Lyme disease is caused by a bacteria called *Borrelia burgdorferi*. Lyme disease, or Borreliosis, is transmitted by the deer tick *Ixodes dammini* and several related species. Additional insect vectors include other hard ticks (e.g., *Dermacentor variabilis*). There are many species of ticks and those that carry Lyme disease are relatively few. On the other hand, these Lyme disease carrying species are widespread. The disease exists along the northeastern states and has been reported in Minnesota, Wisconsin, Canada, Australia and

Ticks and fleas carry disease, and it is important to check your pet's coat every day to control external parasites.

Europe. It can affect people, dogs, cats and many species of domestic and wild animals. Lyme disease is contracted by the bites of infected ticks at any stage in their life cycle and by contact with the urine or blood of infected animals.

The symptoms of Lyme disease are similar in many species, although the chronic form of illness may be less common in people compared to other animals. Lameness and fever are the hallmarks of the infection, but in addition to affecting the limb joints, the bacteria can attack internal organs including the heart and kidneys. Diagnosis of Lyme disease can be difficult because physical symptoms overlap with so many other ailments and because laboratory results can sometimes be inconclusive. Fortunately, it is well controlled with appropriate antibiotics which should be administered for at least three to four consecutive weeks. Flea and tick repellents are helpful to prevent contact with tick carriers. Daily tick checks should be routine to promptly remove ticks because the bacterial transmission occurs only after the tick has been fastened to the skin for several hours (although it may be sooner). A vaccine against Borreliosis is available for dogs through veterinarians and is recommended in geographic areas at risk.

Dogs that have been sprayed by a skunk cannot be deodorized.

There are several ways to dilute the odor left by a frightened skunk. The oldest method to perfume your pungent pet is to bathe the humiliated hound in a tub of tomato juice. The juice is an effective way to neutralize the skunk scent but it is impractical (few of us stock enough juice to partially fill a bathtub) and also expensive.

Fortunately, there are other attractive alternatives. Commercially available products can be purchased over the counter through most veterinarians and pet supply stores. You still need to bathe your dog once, and possibly twice, to eliminate the lingering odor, but you may not appreciate the fuss or mess in your home. Most pet groomers have these products and are happy to apply them for you.

Recently, a simple homemade recipe based upon the chemical composition of skunk spray has been offered as an alternative to rid the perfume from your pungent pet. Its ingredients are inexpensive to make and are commonly kept in most households. The formula calls for one quart of hydrogen peroxide, one-quarter cup of baking soda and one teaspoon of liquid soap (one teaspoon of a mild dish washing liquid is an acceptable substitute). Make sure to rinse your dog's coat thoroughly after the mix is applied.

Finally, if you think your dog has learned a lesson, think again. Most dogs that have been sprayed by skunks will not hesitate to approach another! Apparently, the rewards of interacting with skunks must outweigh the penalty for most dogs.

Petting or brushing against the direction of coat growth will cause aggression

(See chapter 7, "Aggression," section entitled *Petting or brushing a pet against the direction of coat growth will cause aggression.*)

It is natural to avoid punishment. If a pet learns to associate discomfort with grooming, and that aggressiveness prevents grooming, it does not mean that an animal necessarily will be unfriendly under other circumstances. Combing or brushing against the direction of hair growth can be uncomfortable for some pets, but aggression is not the only way they avoid it.

If forcibly restrained, an animal may become aggressive (irritable aggression is frequently triggered by physical discomfort or pain). If this behavior causes the owner to release, the intolerant pet will learn that aggression is a successful way to stop grooming, and this behavior could become more intense with each session.

Grooming should begin at an early age to accustom your cat or dog to that type of human manipulation. Every attempt should be made to make all grooming experiences as enjoyable as possible:

- To begin, make grooming brief.

- Choose a soft brush that is appropriate for a puppy or kitten coat.

- Reward your young pet with lots of verbal praise, using soft and calm words of reassurance.

- Do not force your pet to endure any interaction that is resented, even if necessary.

- Break the activity into a sequence of short steps each associated with a clearly positive experience, such as food.

- If necessary, give a small food treat for staying immobile even if only during a few brush strokes.(Be sure not to reward for intolerance and struggles to break free.)

Make sure that you are gentle with the brush or comb and that the grooming instruments are clean and appropriate for your pet's coat quality. If you see signs of intolerance to grooming after a fairly predictable time, make sure to stop before your pet tries to escape or warn you. If your pet seems to resent grooming against the direction of hair growth then do not brush that way!

Pets have no dental problems.

Cats and dogs, some more than others, are definitely prone to dental problems. Certain dog breeds, the Yorkshire Terrier and Greyhound for example, and cats such as the Persian, seem to have more difficulties than others.

The leading dental problem in dogs is periodontal (gum) disease. Layers of plaque and tartar can coat the visible dental surface, but can also go under the gum line and cause gingivitis (inflamed gums) and receding gums. Gingivitis, particularly in cats, can also be due to viral infection or immune-related inflammation. Dogs and cats can also suffer from cavities above or below the gum line, where they are more difficult to detect. The signs of dental or periodontal disease in pets parallel those in people. Affected pets can be reluctant to drink very cold water, or hesitate to eat their favorite foods or chew on a special treat or toy.

Pets will usually allow their owners to view their teeth by simply lifting the lips. Many pets resent having their mouths pried open if it is not gently attempted, but if possible, this is helpful if an oral problem is suspected. Gums that appear red, puffy and that bleed easily, as well as bad breath, are important warning signs. Broken or loose teeth, changes in pigment of the oral surface or tongue, and growths along the gum line, lips or tongue should also be brought to your veterinarian's immediate attention.

Concern regarding your pet's healthy mouth is good preventative health care. Oral disease frequently affects other parts of the body. Tooth root abscesses can perforate facial bone, for example, and can even disseminate bacteria to the heart and other vital organs. Keep up with your pet's annual examinations so that your veterinarian can monitor the need for any dental work before small problems turn into big ones.

You can't brush your pet's teeth.

The best way to treat many problems is to prevent them in the first place! So. . .brush your pet's teeth! Do not use your own toothpaste, since most pets won't like the taste and these products can irritate their mouths and stomachs. Purchase a toothpaste intended for pets or use a paste of baking soda mixed with water or broth for flavor.

It is usually easiest to begin brushing when your pet is young, but many older pets will adjust quickly if you are patient and gentle. Begin by softly running your finger over your pet's teeth, gums and the inside of the lips for a few seconds. It can be easier to start just near the front teeth before moving to massage the molars and back of the mouth. Concentrate on the cheek side of the teeth since the "tongue" or inside surfaces are rarely a problem. Do this when he or she is in a passive state, for instance, before a nap. Give your dog or cat a small treat when you are done.

Gradually increase the length of time to a few minutes. Imitate the brushing motion you will eventually use with your "toothbrush." A circular motion over the tooth surface is recommended but also use a linear stroke from just above the gum line to the tip of the tooth and the grooves between teeth. Practice this daily or every other day for about three weeks. Once this is well tolerated, graduate to a dampened soft face cloth or dishtowel or even a piece of pantyhose wrapped around your finger tip. Begin the learning sequence again from the start. After several weeks of careful practice, you and your

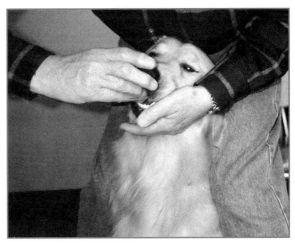

Brushing your pet's teeth can be done with such things as a soft cloth or special pet toothbrushes that fit over your finger. (photo courtesy Maurice Schwartz)

pet are ready to graduate to a soft bristled toothbrush. You might choose one that is intended for pets and fits over your index finger but, if you prefer, you can continue with paste and a soft cloth. Brushing your pet's teeth at least several times weekly can be very helpful in delaying or even preventing the accumulation of tartar that is so injurious to oral structures.

Pets do not get dental cavities (*caries*).

Although these are not as prevalent as they are in people, cavities (also called *caries*) are problems for pets, too. In fact, the most common dental problem among pet cats is hard-to-detect cavities below the gum line. Dogs get cavities, too, but gum disease is more prevalent. By the time dental problems become obvious, simple cavities have progressed to infections of the tooth and surrounding gums. That is why, more and more, preventative dental care is being advocated.

Eating dry food takes care of a pet's basic dental care

(See also chapter 13, "Nutrition," section entitled *Dry food prevents dental problems*.)

Wouldn't it be nice if dental care was as easy as eating a certain food? Unfortunately, for ourselves as well as for our pets, some effort is required to maintain healthy teeth and gums.

The most reliable methods include regular veterinary examinations, maintaining good general health and keeping plaque or tartar from accumulating on teeth surfaces by brushing your pet's teeth. Some studies suggest that pets fed soft food have a higher incidence of dental plaque and tartar compared to those who eat kibble type food.

This evidence, however, is controversial. Most dry foods simply crumble when your pet bites down, for instance, and so provide little scrubbing action on tooth enamel. If you really observe your pet eating dry food, you will realize that most of it is not chewed at all, but is swallowed whole. Thus, limited benefit can be expected from a diet that might even have some negative abrasive value.

Other than the consistency of the food, many additional factors contribute to oral health in cats and dogs. Some pets that eat only dry food have severe tartar buildup, while others that eat soft foods never require dental scaling. Individual genetic predisposition is probably as important or more important than whether the food is moist or dry. Genes underlie breed predisposition (small breed dogs, for instance) and also determine contributing factors like facial structure (such as the flat faces of Persian cats or Pekingese dogs), pH of the mouth and teeth alignment. Rawhide bones may be helpful for dogs that chew them long enough or often enough to derive benefit for their teeth and gums. Occasionally, rawhide bones can cause pasty stools or rarely, intestinal obstruction in overly eager dogs. Feed your pet a quality pet food that it enjoys and teach it to tolerate brushing its teeth from an early age whenever possible.

Nothing can be done to treat a pet's tooth or gum problem.

Don't assume that nothing can be done to treat your pet's halitosis or to prevent tooth loss. There may be many options of which you are not aware that could benefit your pet. Speak to your veterinarian regarding preventative steps that you can follow to preserve your pet's oral health. Veterinary dentists represent a new specialty in modern veterinary practice. Dental procedures currently available to pets include dental scaling, periodontal surgery, extractions, amalgam (fillings), orthodontal procedures and fluoride treatments. Yearly physical examinations by your veterinarian should include inspection of your pet's mouth. Make careful note of any comment and advice regarding your pet's oral health since it may help both you and your pet in the long run.

Chewing bones on a regular basis prevents/treats dental tartar.

(See also chapter 13, "Nutrition," section *Chewing bones is good for your pet.*)
Along with the domestication of animals came our responsibility for their care. Nothing seems more natural than a dog or cat chewing on a bone. Nature, however, is filled with hidden, and not so hidden, dangers. Bone fragments can lodge in the throat to cause respiratory distress. Large pieces can cause intestinal blockage and require emergency surgery. Even small particles of bone can irritate the bowel wall and cause diarrhea or make the stools bulky and cause constipation. Although any bone can splinter, chicken and turkey bones in particular are porous and splinter into daggerlike fragments that slice the mouth and digestive tract.

In theory, regular gnawing on bones sounds like a good thing for your pet's teeth. In practice, the risk of giving bones far outweighs any benefit the pet could derive. Instead, concentrate on maintaining your cat or dog's teeth with regular brushing and routine checkups. If you feel uncomfortable about depriving your pet of this natural, albeit high risk, activity, restrict your choice of bones to very large beef shank bones for dogs. Remove small pieces immediately.

If when trying to retrieve a harmful bone or bone fragment, your pet becomes aggressive and does not permit your approach, distract the dog with an alternative treat. Despite training to minimize possessive aggression, it might be safest to stop giving bones altogether (see chapter 7, "Aggression"). Supervise your pet while it is enjoying a bone, in case of a problem. Rawhide and nylon bones are probably a safer and more effective compromise. For cats, the occasional tip of a chicken wing might be a welcome treat that is usually safely chewed and swallowed, but unlikely to provide any benefit for teeth.

Nail-biting in dogs means it is time for a pedicure.

Some dogs really do bite their nails whether or not their toenails need trimming! This vice, or neurotic activity, is relatively uncommon and may initially develop as exaggerated grooming during periods of anxiety. Many forms of excessive grooming, including nibbling at feet, are thought to be the individual's anxiety-based efforts to calm down. These can persist even when the pet is no longer anxious. Because nail-biting is generally considered an inoffensive activity and is rarely a cause of any self-inflicted injury, this problem rarely requires intervention. An increase in exercise, social interaction and availability of more appropriate and attractive chewable items would be part of the solution.

If a dog's toenails touch the ground, they are too long.

Contact with the ground is not the only reliable criterion for determining whether or not your pet, dogs in particular, needs a pedicure. The shape of a dog's foot and therefore the shape of the toenails, varies with breed and individual. Doberman Pinschers, for instance, have a compact foot with arched toes. Their toenails tend to grow curved and close to the foot. In contrast, Greyhounds have longer feet so their toenails grow with less of a curve and more outward from the foot. Individual gait, that is, how they move, the amount of exercise (which develops muscles in the legs and feet) and where the dog is exercised (grass versus pavement) determines how the animal will wear toenails down. Working nordic breeds like the Alaskan Malamute or Siberian Husky require strong toenails for traction on snow and ice. Traction provided by slightly longer toenails can also benefit frail geriatric dogs that may not be as steady on smooth or slippery surfaces as they once were.

However, if nails are consistently left too long, the animal's pastern (ankle) is weakened and other orthopedic problems can result. Consider these factors, as well as how uncomfortable your pet seems before you trim pointed (unworn) nails, and how happy your pet is afterward!

Excessive self-grooming is a sign that a cat or dog is meticulous.

Although it is true that some pets seem more or less concerned about self-grooming than others of their species or breed, it is unlikely that this has anything to do with vanity or a concern with personal appearance.

The primary function of self-grooming is hygiene, but it is also a reflection of physical and emotional well-being. Grooming can be considered excessive if the animal spends more time engaged in this behavior relative to the apparent need. Grooming should also be viewed problematic if it results in areas of sparse hair or inflamed skin. A pet that is preoccupied with grooming may be suffering from external parasites (fleas, ticks, mites) or contact with an irritating substance or an allergy to a wide variety of inhaled or ingested compounds.

Individuals who are seriously ill frequently stop grooming themselves. However, overgrooming can also be a nonspecific sign of unrecognized medical problems. Excessive grooming can be a neurotic activity related to emotional stress (psychogenic grooming), such as separation anxiety. In dogs, a lick granuloma can develop from chronic inflammation due to continual licking of the front (or rear) feet or legs. In cats, psychogenic licking can be directed almost anywhere thanks to their flexibility. Common feline targets include the tummy, back, feet and legs. If you suspect your pet is grooming excessively, see your veterinarian.

Old dogs can learn new tricks, although it can be more of a challenge to motivate them.

15

Aging Pets

You can't teach an old dog new tricks.

And why not? While it may be true that long-standing patterns are more difficult to modify, middle-aged and older adult dogs are still capable of learning new behaviors. Adults come with a history of routines and behaviors, acquired habits and learned experiences. This does not mean that an old dog cannot learn new tricks, only that an older student may not be as quick to respond to a new challenge. Young animals are a blank slate compared to an adult animal. Stronger motivation for the individual pet, as well as greater persistence and patience by the owner, may be required to initiate the change.

Often an owner may find it simpler to adapt to the pet's old and undesirable pattern as it is in our nature to follow the path of least resistance. This is more a question of the owner's ability to effect changes, not the pet's unwillingness or inability to learn. The owners of aging animals must be prepared to be more patient and may need to provide their pets with more incentive. On the other hand, the cat or dog might adjust to new direction more easily than we anticipate. Your pet could surprise you, and your only disappointment may be in not having tried sooner.

Aging pets do not need exercise, and preventative health care (such as vaccines or heartworm tablets) no longer matters.

Older adult animals are as susceptible as very young animals to contagious diseases. The physical resistance to illness is frequently lower in an aging pet. Nutrition and exercise are important throughout a pet's lifetime. Dogs should be encouraged to exercise in the form of play and daily walks. Cats may require less exercise to maintain good condition, but they still need quality food and regular exercise throughout their lives.

Even if contact with other animals is restricted, an aging pet can contract disease without ever leaving your home or the confines of your backyard. Mosquitoes carrying heartworm larva and fleas carrying tapeworm and a variety of other diseases can enter your home without invitation. Many viruses are airborne (in microscopic droplets of moisture or dust) and are not prevented by the barriers of doors or windows into your home. Also, illness caused by metabolic changes or degenerative states are common in aging individuals.

It is always best to continue annual veterinary checkups throughout your pet's lifetime. This will give you the opportunity to maintain a good relationship with your veterinarian, who will be familiar with your pet's condition over time and can alert you to any health concerns that you may not be able to detect. Speak with your veterinarian about a geriatric screen, which might involve drawing a blood sample to detect degenerative changes associated with age or latent illness in internal organs. A cardiac evaluation might be recommended and might include a radiograph and electrocardiogram. At the very least, you will be better able to catch problems early on and provide your aging pet with appropriate care and attention.

Older pets do not develop behavior problems.

After ten or so years of coexisting with a pet, most of us will admit that we adapted our lifestyles to fit their habits or needs long ago. Undesirable behaviors may persist at tolerable levels for many years. After gentle prodding, most pet owners will reveal at least some aspect of their pet's behavior that is not completely delightful.

Sometimes, preexisting behavior problems may gradually progress until the owner's tolerance reaches a threshold. By the time this critical point is reached, however, the owner-pet bond may be so deteriorated that the owner is unwilling to invest any further time or energy in correcting the problem.

Changes in the owner's lifestyle may also be at the root of pet adjustment problems. Our pets are like mirrors, reflecting our emotional turmoil as well as our psychological well-being. If we undergo stress, the quality and quantity of our interactions with them is impacted. They are superb interpreters of our facial expressions and body language, sometimes communicating to them much more than we can verbalize to another person.

The owner of an aging animal should be aware that changes in an animal's behavior are often the first indications of underlying medical illness. Urinary incontinence in an aging dog, for example, is a common problem. It may be due to infection or inflammation of the urinary bladder, hormonal imbalances, neurological degeneration of urinary sphincters, diabetes and kidney disease. It might also be due to painful arthritic joints that decrease a pet's mobility and prevent voiding outdoors. Breaks in house training in older pets can be a symptom of separation anxiety, even when it was never a problem before. Older pets can be more sensitive to being left alone. Most of these physical conditions respond to medical intervention. When the possibility of an underlying medical problem has been investigated, behavioral problems in older pets are often responsive to appropriate treatments. Finally, cognitive changes associated with age may affect a pet's behavior, just as organic brain diseases alter behavior in geriatric human patients. Ask your veterinarian to refer you to a veterinary behaviorist in your area.

Older pets do not become senile.

Cognitive change refers to variations in an individual's ability to perceive, interpret and respond to the world. In aging people, it was once assumed that diminished interest in previously enjoyed activities, lowered states of awareness or periods of confusion and withdrawal were inevitable. More and more, emotional and physical deterioration associated with age can be attributed to recognized diseases. Some clinical conditions can be expected to respond to treatment while others remain, for now, therapeutic challenges. Recently, exciting parallels in age-related cognitive changes between people and companion animals have gained attention. While much work remains to be done in both human and veterinary medicine, it is not unreasonable even at this stage of research to draw some comparisons. For the time being, diagnostic terms like senility and Alzheimer's disease apply only to people. Other terms, like organic brain disease or central nervous system deterioration, seem applicable to pets as well.

We have just begun to appreciate some of the physical, emotional and behavioral changes that result from age-related changes. Until more is understood, at least two fundamental statements are indicated:

- First, never assume that your aging pet cannot be treated for any medical or behavioral problem. Annual veterinary visits (more often, if required) throughout your pet's lifetime provide the best opportunity for ongoing evaluation and communication.

- Second, the aging pet deserves no less love and attention than in its younger days. Whether your geriatric dog or cat has problems that demand your additional attention or not, they should not be taken for granted. Life is finite for us all. It must, therefore, be cherished.

A year in a dog's or cat's life is equal to seven human years.

This used to be a popular way to measure a pet's maturity, but it is no longer valid. Recent advances in veterinary medicine have prolonged the anticipated life span of our pets. Thanks to the efforts of veterinary organizations and the pet care industry, pet owners are more informed and better exposed to advice intended to improve the care of their pets.

Today, the scale between dog/cat years and human years can no longer be estimated by a simple equation. It is best measured on a logarithmic scale. The first three years of a dog's life are still equivalent to about seven years each, although some would argue it might be closer to ten. After the age of three, each pet's year is comparable to progressively fewer than seven years. Over the age of ten years, for example, each year in a dog or cat's lifetime would be equivalent to between two and five years for humans. Because the average cat lives longer that the average dog, the scale is even more gradual for cats.

Pets only live about ten years.

Today, the longevity of most pets can be expected to exceed the ten-year mark, thanks in part to advances in pet nutrition, veterinary medicine and owner awareness. The average pet dog can live about twelve or thirteen years. Small dogs, such as miniature or Toy breeds, frequently live fifteen, sixteen years or more. In general, the larger the dog, the shorter the expected life span. Giant breeds such as the noble Great Dane, for example, rarely survive past the age of ten.

Cats do not show the same range of sizes as dogs, so it is far easier to generalize about their anticipated life span. Indoor cats usually live longer than cats that are permitted to roam outdoors. The average indoor cat will likely approach or exceed fifteen years of age. Cats close to twenty years old are less and less unusual. Cats that roam outdoors are exposed to many dangers that adversely affect their longevity.

In addition to lifelong veterinary care, our pets deserve adequate nutrition and lots of positive attention from us owners to help them to live the longest and best quality lives possible.

A pet that has gone blind or deaf cannot adapt and should be euthanized.

There are a multitude of medical problems that cause a pet to lose the ability to see or hear. Blindness and deafness can be genetic, congenital (from birth)

Thanks to the advances in veterinary medicine and dedicated pet owners, the average dog will live about twelve years or more. (photo courtesy of Estelle Schwartz)

or acquired by illness or accident. Sudden and obvious health problems require immediate attention, particularly if they affect delicate tissues of the eye and ear.

Most diseases that affect a pet's vision and hearing, however, happen very gradually. This may be particularly true in aging animals. By the time any deficit is noticed by the owners, pets have had time to adjust to progressive physical changes in their bodies and to adjust behavior accordingly. Medical treatment and long-term management for your older pet's condition is likely available. Report any sudden or gradual deterioration in your aging pet to your veterinarian.

In general, little additional effort is required to care for a pet that cannot no longer see or hear well. In fact, the less interference with set routines and environment, the better. Keep the furniture arranged in the same way so that a sight-impaired pet can maneuver through familiar paths in your home. You may want to place gates in front of stairways in case your pet loses balance or becomes confused. For the hearing-impaired pet, the only additional necessity is your patience. Your pet may not hear you call, for example, so you may need to get attention by visual means. Hearing loss may affect only a limited range of sound or your pet's hearing may be relatively unaffected if sounds are close by.

Experiment with your pet's new physical restrictions to see what may still be effective to get its attention. If your aging pet has been allowed to roam freely outdoors, of course, any deterioration in its ability to perceive or escape danger should convince you to accompany it on future outings or to prevent roaming altogether.

A dog or cat that fails to respond must be deaf.

Many pet owners attribute a pet's decrease in alertness to deafness. Aging affects many physical functions and hearing is certainly among them, even in companion animals. Failure to respond when called, however, is not a reliable test of deafness.

Some pets, particularly as they age, sleep more soundly and may not awaken as rapidly. Their hearing is still functional, but their vigilance might be less than it was in younger days. Some pets were never reliably trained to "Come" in response to hearing their name. Unreliable training should not be confused with hearing impairment. Older pets frequently suffer from arthritic joint disease and may be less able or willing to respond to someone calling them. Finally, deafness in older pets may affect only a range of audible sound. A hearing-impaired pet may be deaf to a voice calling from a distance, but respond to other louder noises close by or within range of their individual sensitivity.

Aging cats do not suffer from arthritis.

Age-related arthritis affects cats as well as dogs, but degenerative joint disease can have many causes in both dogs and cats.

There are many more congenital orthopedic problems in dogs than in cats, but they are equally prone to traumatic bone or joint injury. Cats can suffer fractured limbs and dislocated bones as a result of being struck by cars, for example. Because cats generally have a lighter body frame than do dogs, the effects of arthritis may not be as obvious. The same degree of arthritic change in the spine or hips may not bring about extreme physical dysfunction in a ten-pound cat as it might in a ninety-pound dog. Arthritic changes related to old injuries may not appear for years. Aging cats can indeed suffer from arthritis, but may accommodate these changes better than the average dog.

Increased thirst and more frequent urination are normal in older pets.

Water intake and urinary output are balanced in the healthy individual according to activity and other daily demands. Changes in thirst may fluctuate with ambient temperature, for example.

Age does not necessarily bring an increase in thirst and urine production and this is not normal in the older dog or cat. In fact, this symptom is common to many medical disorders and should alert the owner's concern. However, some of the physical problems associated with drinking and

Sara, almost fifteen years of age, is playful and alert, keeping vigil while her owner works.

urinating more frequently are seen more often in middle-aged or older pets. Degenerating kidney or liver function, diabetes and infections of the urinary tract and bladder can all affect fluid balance. Medication, diet change or anxiety, perhaps related to the addition of new pets or other household members, can all heighten thirst. It is important to understand, however, that many of these problems can be seen in younger animals and that many other details are needed to confirm the diagnosis of either a medical or behavioral problem.

Incontinence (urine and/or stool) is normal in aging pets.

The control of urinary and anal sphincters is a complex of physiological mechanisms. Failure of any of these control mechanisms can lead to involuntary passage of urine or stool. There are many metabolic illnesses, infectious diseases, neurological and muscular disorders, and medications that can affect sphincter control. Some of these conditions, such as bladder infections, occur in animals of any age, but specific physical ailments can be more common to a given age group. Incontinence occurs more in aging pets because of neuromuscular deterioration, for example, but this does not make it normal or acceptable.

The most common causes of urinary incontinence include bacterial infections of the genital or urinary tract. Appropriate antibiotics are usually curative. Medications can also affect sphincter mechanisms. For example, corticosteroid anti-inflammatory drugs used to control allergy can cause excessive thirst and therefore increased urine production that overwhelms bladder control. Urinary incontinence can be related to hormonal imbalances. This is more common in neutered dogs of either sex, but is more frequent in spayed females. Treatment may include hormone replacement or other medications which improve sphincter tone. This type of incontinence is rare in cats. Degenerative neuromuscular disorders directly impact the nerves and/ or muscles composing the sphincters. This group of conditions can be somewhat more resistant to treatment.

House soiling in aging or debilitated pets can be very distressing to both pets and owners. Always seek professional help as soon as any problem arises. Early diagnosis usually means that available and appropriate treatment can begin immediately, too.

16

Pet Death

The best death your pet can have is to die at home without interference.

Asked to define the criteria of a "good death," most of us would include a lack of pain or discomfort, a desire to be surrounded by loved ones and the option of dying at home. We all secretly wish for our pets to pass peacefully at home, slipping silently into sleep.

Euthanasia, an option available for pets, is at once a blessing and a burden. The decision to end a life is a heavy one for all involved, but the greatest emotional repercussion lies with the pet owner. When circumstances are such that it is unbearable or unthinkable to prolong the life of a sick or dying pet, euthanasia becomes an opportunity to provide relief to both pet and, therefore, to the owner. For some, euthanasia is simply not an option under any circumstances.

For pet owners who view euthanasia as an acceptable alternative to inevitable and terminal suffering, a compromise of sorts may be available. To fulfill the wish of dying at home, some veterinarians still make house calls to perform these merciful procedures. This may be ideal for a pet difficult to transport or one with a history of fearful reactions at the veterinary clinic. The owner is also saved from arranging transportation and is spared embarrassment with displays of emotion in an intensely private time.

If this cannot be arranged in your area, bringing your pet to the clinic for this final act of kindness still has comforting advantages. You will be surrounded by compassionate people who share your love of animals. You will not need to be concerned with housecleaning if your pet soils, as sometimes happens when sphincters release in death. You will be able to return to a house that is filled with echoes of a living pet and not with the memory of final minutes. Wherever euthanasia is performed, the most important thing is your success in relieving the suffering of someone you love.

Pets suffer when they are euthanized.

Euthanasia, meaning "good death," refers to everyone's desire for a painless and easy transition into lifelessness. Euthanasia may be selected as the kindest and best treatment option for an ailing pet after thorough discussion and support by your veterinarian. Your pet will likely be held by a veterinary technician to assist your veterinarian in the intravenous injection of a concentrated and lethal dose of anesthetic. A slight pinch might cause a moment's discomfort as the needle tip pierces the skin, but the discomfort is no greater than any other injection. As soon as the needle enters the vein (usually in the pet's front leg although other blood vessels may be selected), the euthanasia solution is injected. Unconsciousness occurs almost immediately and your pet's eyes will likely remain open. As the entire contents of the syringe are administered, breathing and heartbeat slow to an end. The entire procedure may last only seconds. Your veterinarian will verify cardiac arrest with a stethoscope. If you remain with your pet to say your final good-byes, do not be alarmed by any muscle motion. In the first few moments of death, muscles can contract involuntarily and cause twitches of the skin and diaphragm, triggering movements that resemble respiration. Urine and stool may also be voided as sphincters relax. These are expected and do not mean that the injection failed.

Despite your grief, memorize the feel of the soft coat, the sweet face, the lifetime of mischief and devotion. Be grateful that your life was touched by theirs. Be gratified that you could provide the final gift of, if not a "good" death, then at least a better one.

Mourning the loss of a cherished pet is not the same as grieving for a person.

To a devoted pet owner, the death of a cherished pet can be as shattering as the death of a friend or family member. Indeed, the loss of a pet could even impact us more. Our dogs and cats live with us every day, sharing in our daily burdens and offering continual, unconditional support and companionship. In a world where many of us lead solitary lives, our pets become an integral part of our lives. Their significance can surpass our relationships with many other people.

Losses are part of life for all of us. We move from one town to another, to a new school, a new job. We make new friends along the way and they leave

My first childhood pet, a Samoyed, affected the course of my life.

us or we leave them. Sometimes we have no choice but to let people and places go, and other times we choose to cut the bonds. Death is the only permanent loss. The loss of something of value, regardless of whether it is a person, a pet, an object or location, is a loss, nonetheless. All losses lead to grief. The subject of our grief is not as important as its value to us. The intensity of grief is determined by the significance of what or whom we lose. The process of grieving is always the same. You cannot feel the depths of my grief and I cannot feel yours, but if I put aside the differences between us I will catch a glimpse of your despair. If I focus only on the similarities between us, then I need not judge whether or not you should mourn, only that you do.

Pet owners who do not cry when their pets die do not care.

Grieving the loss of a pet is a time of distress for most pet owners even when they seem to be unaffected. We should not be quick to project our own perception or expectation of someone else's behavior.

As individuals, we love in different ways. Similarly, our public expression of grief reflects our individuality and the unique relationship we have with a pet. Our pets represent many symbolic roles, unknowingly taking the part of a supportive friend, attentive parent or surrogate child. Regardless of the grief we may or may not show, the emotional process of mourning a pet brings us through steps which are common to us all. There are five phases of mourning which we all experience, regardless of how openly we display our pain or whether the one we mourn walked on two legs, four legs or no legs at all:

1. The immediate reaction to death is denial and a need for isolation. This is a normal defense mechanism that buffers the shock and gives us time to mobilize our physical and emotional energies.

2. When the first wave of pain begins to subside, feelings of anger may swell. This is the emotional backlash to the initial paralyzing effect of grief. We may be angry at anyone that crosses our path. We may over-react to inconsequential events. We may be angry at the veterinarian who delivered the news of a terminal illness or relieved the pet of suf-fering. We may be angry at the pet for leaving us. We are angry in or-der to compensate for feelings of regret, guilt or vulnerability. Death reminds us that we have no real control over much of what life brings.

3. Helplessness is a frightening thing. And so the next phase of mourning is an effort to regain a sense of control. We enter a stage of intellectu-alized emotional gymnastics referred to as the bargaining phase. If only we had not waited so long before seeking veterinary care. If only we had left work sooner or were not detained by traffic. Bargaining with a higher power is common at this time. If only our pet might survive we would surely attend religious services more often. As feelings of vul-nerability and mortality persist, depression deepens.

4. The depression associated with mourning has two distinct but overlap-ping sources. A deeper sense of depression relates to our most intimate preparation to say good-bye to a cherished companion. This is distin-guished from the practical aspects of separation that depress us in more obvious ways. Morbid decisions regarding when and how to terminate treatment, burial arrangements, anticipated bills, and a recognition of how empty the house will be, contribute to a different type of depression that coexists with the first. Depression is a normal reaction when the demands of life (and of death) become overwhelming. However, feelings of complete hopelessness, isolation and thoughts of self-destruction should be warning signs to you, and to friends and family.

5. At some point in time, a point in time which is decidedly different for each of us, death is normally accepted. Accepting the loss of a loved one is the resolution phase. It may be reached long after the death has occurred and sometimes never quite achieved. For those who are fortunate to have advance warning and enough time to grieve, this phase may even be reached prior to the death.

These five phases of mourning do not necessarily follow in sequence. Frequently, the stages concur, and we may experience regression to and repetition of each step. The expression of grief is also determined by the culture and family in which we were raised. The fundamental source of strength for us all, regardless of the phase of grief or our family background, is hope. Despite our profound losses there remains the spirit of life and of hope for better times to come.

When your pet becomes ill or dies, your friends, coworkers and family will understand your loss.

The emotional bonds with our pets can often be difficult for anyone else to comprehend. When a pet becomes ill or dies, other animal lovers, at least, can honestly empathize. For people who find no particular affinity for the rest of the animal kingdom, it can be difficult and even impossible to understand the loss of a pet. Yet, it is often in times of crisis that we reveal ourselves to each other. Friends that you thought would surely stand by you may be unable or unwilling to provide you with any measure of comfort. Conversely, coworkers that seemed oblivious to your existence and even total strangers may sincerely volunteer to aid you when you are most vulnerable.

The reality is that society does not value the life of a dog or cat (or anything else) as much as it values a human life. It is natural, and indeed desirable for the survival of our species, to instinctively favor the existence of any creature whose genes most closely resemble our own. Some people are interested only in the genes of human beings and are unable to appreciate the intrinsic value of any other form of life. It is unrealistic to expect them to comprehend that, over the course of a lifetime, a four-legged animal has become intrinsic to our lives. Recognize the limitations in others and avoid those people who cannot feel for you, or worse, seem to amplify your grief. The process of mourning a pet is a stressful time and emotional healing demands much concentration. Protect and pamper yourself. Surround yourself with a support network of people who, at some level, understand.

Children do not understand when a pet becomes ill or dies.

Children begin processing concepts of illness, mortality and death from an early age. Their thoughts may be subconscious, and they may not be able to verbalize many of them until later in life. Many childhood games, such as "hide and go seek," and a common childhood fear of the "bogeyman," may symbolize early interpretations of death.

Young children may have difficulty understanding the finality of the loss of a pet but they will, at the very least, be aware of the absence. They will also most certainly respond to the grief they sense in their parents. Children respond best to honest and direct explanations. Reassure your children that every effort is being made to make their pet well again. Let them know that you and the veterinarian are doing everything possible so that their pet can come home safe and sound just as before. If a pet is dying, a simple statement to prepare them for their loss is appropriate. Suggest to them that, even though you all wish things could be different, life is ended at some point for all living things. Sometimes all that is needed after that is a hug.

This is not to say that the experience of death should be imposed upon a child. It seems most sensible and sensitive to allow the child to decide how much more he or she needs to see or wants to find out. If a pet is brought to the veterinary clinic for euthanasia, a young child need not be exposed to the procedure. Indeed, many adults do not elect to attend. Children should not be raised in an insulated bubble, but neither should they be traumatized. If a pet has died, it is often helpful to conduct a memorial service so that the child has the opportunity to express deeper emotions and at the same time be comforted by a sense of hope and closure.

To explain the death of a pet to a young child, parents should use the terms "put to sleep" or "went to heaven."

Every parent has a natural desire to protect their child. The role of a parent, however, should also be that of a teacher. The death of a pet is a valuable opportunity to teach a child about life. We should not let our own fear of death interfere with acknowledging the grief of a child. Parental attempts to camouflage death can be more damaging than beneficial. Phrases referring to the euthanasia of a pet, such as "putting to sleep," are entirely inappropriate. Some children develop a fear of bedtime or become terrified of falling asleep for fear they will not awaken. A child that is told that their pet "went to heaven" might become angry at God for depriving them of their friend and playmate. Adults can be unaware that children can respond so literally to what they hear.

Children begin incorporating concepts of death from a surprisingly young age. Their courage and ability to understand should not be underestimated. Children respond best to honesty. Intentionally misinforming a child about the death of a pet might only complicate perceptions later. Some children carry with them a sense of doubt and mistrust when parents tell them that the dog was given away to a farm or that the cat decided to live somewhere else.

The death of a pet is an opportunity for parents to encourage a child's ability to communicate emotions. It is an opportunity to deepen parental bonds as the child is reassured by comforting adults. Losses are inevitable in life. Our children deserve to be prepared for all of life's experiences.

You must be present at the time of your pet's euthanasia.

Watching the death of a loved one is perhaps the most stressful, heart-wrenching and emotionally powerful experience anyone can endure. It is no less intense an effort for a pet owner. The decision to accelerate the inevitable is a tremendous responsibility accompanied by self-doubt and guilt. Although euthanasia is painless and swift, it is often overwhelming for the pet owner.

Intense grief often causes physical symptoms such as shortness of breath and dizziness. Pet owners with underlying medical problems should consider their own health as much as that of their beloved pets. Pets are sensitive to your emotional state and some might be calmer if you were not present. No one should feel obligated to attend the euthanasia of a pet. Veterinarians are extraordinarily gentle during this procedure. It can be easier for pet owners to wait in an adjoining room and to say their last good-byes once it is over. Some pet owners find it best for them to send a friend in their place or to say good-bye and leave their pet before the euthanasia is performed. Many people prefer to remember when their pet was alive and avoid the memory of seeing death.

For those who feel that they must attend, perhaps out of loyalty, a sense of duty or devoted friendship to the end, a sense of closure may not be otherwise possible. The choice of whether to remain or not is as individual as our own spiritual beliefs and our emotional response to death. Do not let anyone make that decision for you in case you have regrets later. No one but you can say what is best for you in this situation.

Think about it and discuss it with your veterinarian. Do what is best for your pet, but do what is best for you, too. If you are physically and emotionally able, you might find comfort in knowing that your pet's last sensations included you. Whether or not you are present, what is ultimately more important is the quality of your pet's lifetime and your memories of happier times.

You should get another pet immediately to replace the one you lost.

We experience grief in many ways. When we lose a pet, the grief we feel will be influenced by our past experience with death, by the quality of the relationship with the pet, by the circumstances of death and by the circumstances of our own lives at the time. The mourning period is a natural reaction to loss and an essential time to recuperate. It is an opportunity to reflect not only upon the pet we mourn, but on other losses that have yet to heal entirely. Life crises have a way of exposing psychological events that were insufficiently processed. Grieving a pet can be a valuable opportunity to nurse other emotional wounds that are incompletely healed, to reach out to other people that deserve greater appreciation, to project into the future and focus on dreams that remain to be fulfilled thus far.

Sometimes, replacing a pet with a new one interferes with important reflection and insight. Sometimes, it is better to deal with old business before taking on new challenges. Of course, a new pet can be a welcome distraction, filling the emotional void and diluting our grief. For some of us, this is necessary to make the pain more bearable.

Do what you need to recover from your loss. We all heal in different ways and should be allowed to do so. If you would prefer to take the time to travel rather than to house-train a new puppy, you should. If the companionship of a new pet is a priority, than you should not feel guilty or disloyal to the memory of the one you grieve. Remember that your new pet will never be the same as any predecessor. The newcomer's talents, intelligence and unique characteristics should not be compared to your previous pets'. Every pet is a lifetime investment. If you cannot welcome your new pet with the time, attention and affection it deserves then, indeed, you are not ready and should wait until you feel you are.

Veterinarians perform experiments on dead pets.

Veterinarians engaged in private practice deal with the care of healthy and sick pets all day. As doctors of veterinary medicine, they are trained to diagnose disease, to recommend preventative steps in health care, to offer advice regarding a patient's well-being and to explain options for treating injured or sick pets.

Veterinarians in private practice do not dissect or abuse the bodies of pets that have been euthanized or that died in their care. Pets, whether alive or dead, are the property of their owners. Any postmortem examination is forbidden unless expressly consented to by the pet's owner. When a pet has died, pet owners discuss burial options with their veterinarians, who then comply with owners' wishes.

Veterinarians are unaffected by performing euthanasia because they do it all the time.

Veterinarians choose this challenging profession for many individual reasons and to pursue a variety of goals. The most common motivation for an interest in the study of veterinary medicine is a love of animals. The most important ongoing source of gratification is the intention of most veterinarians to do what is best for their patients.

Veterinarians strive to relieve the suffering of patients while considering the animal's quality of life beyond immediate medical care. Many veterinarians have the opportunity to follow their patients for years, watching them grow from chubby puppies and fuzzy kittens through adulthood and into old age. It is frequently impossible not to develop a certain affection for many of our patients. Professional relationships with pet owners deepen with time, confidences are exchanged and an alliance formed to ensure the pet's physical and emotional well-being.

Veterinarians cannot help but be touched by the loss of a patient. Certainly some cases seem to affect us more than others, but the empathy for a pet owner's grief is constant. Veterinarians are individuals, each with our own psychological baggage and degree of comfort in emotionally intense situations. We may be more or less equipped to satisfy a pet owner's need for a simple hug or to offer reassurance in times of critical decisions. Not all veterinarians allow their professional guard to come down in the presence of a grieving owner. Indeed, it is generally not considered professional to do so and often would not be helpful to the grieving pet owner. Most of us, therefore, rationalize our immediate emotional response by reaffirming that our action is in the best interest of the pet and the owner. Our compassion must not affect our objectivity. Veterinarians carefully think over every decision to euthanize a pet. Every case is different and all options should be explored, including the quality of life for both pet and owner.

If a veterinarian objects in any way to a request for euthanasia, they are not obligated to do it and may refer the owner elsewhere for a second opinion. No one enjoys ending a life. Veterinarians simply have the privilege of offering an alternative solution when a patient's suffering is intolerable or incurable. Euthanasia can be a blessing for the pet and the owner, but a good veterinarian is humbled with every one.

Index

Abyssinian cats, 113
Acquired immune deficiency syndrome (AIDS), 176
Afghan Hounds, 50–51
Aggression, 79–101. *See also* specific behavior
  in American Pit Bull Terriers, 90
  in black cats, 19
  in calico cats, 98
  competitive, 94
  declawing and, 98–99
  disturbances during feeding and, 93–95
  disturbances during sleep and, 79
  going against direction of coat and, 211–12
  toward infants, 62–64, 65–66
  toward men, 93
  neutering and, 144–45
  between newly introduced cats, 42–43
  between newly introduced dogs, 41–42
  toward owners, 79–80, 96
  toward owner's mate, 96
  possessive, 94–95, 136
  toward postal carriers, 83–84
  rabies and, 97–98
  rough play and, 82–83
  seizures and, 96–97
  in small dogs, 45–46
  staring and, 87–89
  territorial, 94, 135–136
  in veterinary clinic, 91–93
Aging, 219–26
Akitas, 100
Allergies
  to food, 198–99
  nose pigment and, 4–5
  to pets, 52–53
  sneezing and, 173–74
Alpha roll, 71

American Pit Bull Terriers, 25, 90
Anogenital investigation, 156–57
Apartments, large dogs and, 44–45
Arthritis, 224
Attention span, 7
Australian Shepherds, 3, 75

Bad breath, 177
Barking, 149–50
Bathing, 203–4
Biting, 79–80, 81–82, 89–90, 167–68
Black cats, 19
Blindness, 204, 222–23
Boarding kennels, 159–60, 161–62
Body odor, 203
Bones, 190–92, 215–16
Border Collies, 50
Boredom, 108, 193–94
Breeders, 55
Brewer's yeast, 194
Brucellosis, 146
Brushing and combing, 204–5, 207–8, 211–12
Burmese cats, 53

Calico cats, 17, 98
Car sickness, 159
Car travel, 26, 30–31
Cat fights, 99–100
Catnip, 11–12, 110–11
Cats, 9–19. *See also* Female cats; Male cats; specific breeds
  ability to land on feet, 16
  AIDS and, 176
  behavior of newly introduced, 42–43
  bonding by, 1–2
  busy people and, 47–48
  chewing by, 113–14

Cats, (cont.)
  color vision of, 4
  coprophagia absent in, 121–22
  demonstrative behavior in, 9–10
  digging by, 104–5
  dogs and, 100, 101
  ear hair in, 17
  failure to cover waste, 128–29
  heartworm in, 171
  as hunters, 12–13
  hyperactivity in, 153
  indoor confinement of, 13, 14–15
  milk and, 192–93
  night vision of, 16
  obedience training for, 75–77
  outdoor dining by, 185
  possessive aggression in, 95
  purring in, 10
  ringworm and, 175–76
  scratching of surfaces by, 104, 108–9,
    112–13
  tail wagging in, 10–11
  tongues of, 206–7
  Toxoplasma and, 61–62
  urine marking by, 120, 127–28
  water intake by, 192–93
  as women's pets, 46–47
Cat toys, 158–59
Chewing, 113–14
Children, 57–66
  death of pet and, 232–33
Chocolate Labrador Retrievers, 51
Choke collars, 70–71
Chows, 100
Claws. See also Declawing
  biting of, 216
  proper length of, 216–17
  trimming of, 108, 112–13
Coat, 195, 197
Colds, 174
Cold weather, 31
Collies, 3
Color vision, 4
Competitive aggression, 94
Constipation, 197
Cooking oil, 197
Coprophagia, 116, 121–23
Crate training, 151–52

Dalmatians, 3, 75
Deafness, 3–4, 75, 222–23, 224
Death, 227–36. See also Euthanasia
Declawing, 98–99, 108, 111–12
Dental problems, 177, 189–90, 212–16
Destructiveness, 103–14. See also specific
    behavior
Diabetes, 124, 127–28, 192, 225
Diarrhea, 124, 168–69

Digging, 103–5
Distemper vaccine, 167–68
Doberman Pinschers, 25
Dogfights, 99–100
Dogs, 21–32. See also Female dogs; Male
    dogs; specific breeds
  AKC registration papers for, 27
  ancestry of, 21
  barking in, 149–50
  behavior during transitional toddler phase,
    64–65
  behavior of newly introduced, 41–42
  bonding by, 1–2
  cats and, 100, 101
  cold weather and, 31
  color vision of, 4
  coprophagia in, 121, 122–23
  crate training for, 151–52
  deodorizing after skunk spray, 210–11
  digging by, 103–4
  dominant paw in, 29–30
  ear hair in, 171–72
  eyes, hair in, 204
  fenced yards for, 21–25, 44
  fighting in children and, 64
  growling in, 89–90
  "guilt" after accidental soiling, 117
  howling in, 149
  hyperactivity in, 153–54
  intelligence of, 49–51
  jumping on people by, 158
  large, 44–45
  need to please humans, 32
  proper length of toenails in, 216–17
  rolling over by, 152–53
  running away by, 27–29, 116
  small, 45–46, 74–75
  staring at, 87–89
  tail wagging in, 89
  teaching new tricks to old, 219
  urine leakage in, 119–20
  urine marking by, 120
  walking of, 21–22, 116
Dog toys, 158–59
Dominance behavior, 86, 94–95
Dreaming, 2–3
Dry pet food, 185–86, 189–90, 214–15

Ear cropping, 172
Ear hair, 17, 171–72
Elimination problems, 115–30. See also
    specific problems
Estrus (heat), 131–32, 140–42
Euthanasia, 227–28, 233–34, 235–36
Exercise, for aging pets, 219–20
Eyes
  brown discharge from, 174
  hair in, 204

Feeding. *See* Nutrition/feeding
Feline leukemia virus (FeLV), 146, 176
Feline urinary tract disease (FUTD; feline
        urological syndrome), 129–30, 145,
        185–86
Female cats
    frequency of estrus (heat) in, 142
    instinctive care of offspring, 133
    mating preferences in, 138
    neutering of, 131–32, 137
    obesity in, 183
Female dogs
    instinctive care of offspring, 133
    mating preferences in, 138
    menstruation in, 140–42
    mounting behavior in, 86–87
    neutering of, 131–32, 135–36, 137
    urination by, 118
    as watchdogs, 80–81
Fenced yards, 21–25, 44
Fish, 181, 195
Flea collars, 208
Flea pills, 209
Fleas, 194, 208
Food rewards, 69–70
Free-feeding, 181–82, 184

Garlic, 194
Gender. *See* Female cats; Female dogs; Male
        cats; Male dogs
German Shepherds, 90
Golden Retrievers, 91
Grass eating, 173, 199–200
Grooming, 203–17. *See also* specific grooming
        measures
Growling, 89–90

Hairballs, 204–6
Health, 163–79. *See also* specific disorders
Heartworm, 170–71
Heat. *See* Estrus
Hot weather, car travel in, 30–31
Howling, 149
Hunting dogs, 37–38
Hyperactivity, 153–54

Incontinence, 124, 220–21, 225–26
Infants, pet aggression toward, 62–64, 65–66
Intelligence, 48–51
Intestinal parasites, 124, 168–69, 193

Jealousy, 96

Licking, for healing, 164–65
Life-span, 222
Litter boxes, 61–62, 104, 125, 128–29, 186
Loneliness, 38–39
Long-haired cats, 205–6

Loud noises, 154–55
Lyme disease, 209–10

Maine Coon, 6
Male cats
    dry food and, 185–86
    ear hair and, 17
    first time mating by, 135
    neutering of, 132–33, 136–37, 139–40
    nipples of, 145–46
    obesity in, 183
    offspring recognition by, 134–35
Male dogs
    behavior toward infants and, 65–66
    first time mating by, 135
    neutering of, 132–33, 136–37, 139–40
    nipples of, 145–46
    offspring recognition by, 134–35
    urination by, 118–19
Mammary tumors, 131, 132
Mange, 175
Masturbation, 136, 140, 146–47
Mating, 135, 138–39
Meat, 195–96
Medication, 153–54, 164
Men
    aggression toward, 93
    fear of, 155–56
Menstruation, 140–42
Milk, 192–93
Mounting behavior, 84–87
Mutts, 25–26

Nails. *See* Claws
Name of pet, changing, 5–6
Neutering
    aggression and, 144–45
    behavior changes following, 139–40
    effect on happiness, 132–33
    before first estrus (heat) or litter, 131–32
    masturbation and, 136, 140, 146–47
    mean behavior and, 144
    mounting behavior and, 86
    obedience training following, 67–69
    obesity and laziness following, 142–44
    organs removed in, 137
    protectiveness of dogs and, 135–36
    urinary tract infections and, 136, 145
    urine marking and, 127–28
    vasectomy compared with, 136–37
Night vision, 16
Nine lives myth, 9
Nipples, 145–46
Nose
    as health indicator, 165–66
    pigment of, 4–5
Nutrition/feeding, 115–16, 181–201

Obedience training, 4, 67–78. *See also* Punishment
Obesity, 142–44, 181, 183, 184, 186
Offspring, 133, 134–35
Old English Sheepdogs, 75

Pack defense, 136
Panleucopenia, 167
Paper training, 115–16
Paw size, as predictor of growth, 6–7
Persian cats, 53
Pet foods, 185–86, 187–90, 214–15
Pet selection, 33–55
Pet stores, 55
Pick of the litter, 35–37
Pick-up trucks, riding in back of, 26
Play-biting, 81–82
Possessive aggression, 94–95, 136
Postal carriers, 83–84
Pregnancy
    contact with cats during, 61–62
    false, 137–38
Psychogenic hyperphagia, 182
Punishment, 29, 70–72, 116–17. *See also* Obedience training
Purebred dogs, 25–26
Purring, 10

Rabies, 97–98, 163
Rabies vaccine, 166, 168
Raw eggs, 195
Raw fish, 195
Rewards, 69–70
Rex cats, 53
Ringworm, 163, 175–76
Rolling over, 152–53
Rottweilers, 25
Running away, 27–29, 116
Runt of the litter, 37

Salivation, 37–38, 200–201
Samoyeds, 75
Scooting, 124
Scratching, of self, 208
Scratching, of surfaces, 104, 108–9, 112–13
Scratch posts, 99, 108, 109, 110–11
Second pets, 38–43
Sedation, for travel, 150–51
Seizures, 96–97

Senility, 221
Separation anxiety, 105, 107–8, 124, 125, 221
Sexually transmitted diseases, 146
Shedding, 197–98
Sheepdogs, 51–52
Shetland Sheepdog, 3
Shock collars, 24–25, 71
Short-haired pets, 207–8
Siamese cats, 48–49, 53, 113
Skunk spray, 210–11
Sneezing, 173–74
Staring, at an aggressive animal, 87–89
Starvation, elimination problems and, 123–24
Striped markings, 17, 18

Tabby cats, 18
Table scraps, 194
Tails
    docked, 172
    striped, 17
Tail wagging, 10–11, 89
Tapeworm, 209
Territorial aggression, 94, 135–36
Thiamine deficiency, 195
Thyroid gland hyperactivity, 153
Ticks, 209–10
Tongues, of cats, 206–7
Toxoplasma, 61–62
Travel, sedation for, 150–51

Urinary tract
    infections, 124, 136, 145. *See also* Feline urinary tract disease
Urination frequency, 224–25
Urine leakage, 119–20
Urine marking, 120, 127–28

Vaccinations, 166–68
Vegetarian diets, 187
Veterinarians, 177–78, 234–36
Veterinary medicine, 178–79
Vocalization, by Siamese cats, 48–49
Vomiting, 168–69

Watchdogs, 80–81, 135–36
Water intake, 126–27, 192–93, 224–25
Whisker trimming, 17